Editors
Lorin Klistoff, M.A.
Kathleen "Casey" Petersen

Managing Editor
Karen Goldfluss, M.S. Ed.

Illustrator
Kevin McCarthy

Cover Artist
Brenda DiAntonis

Art Manager
Kevin Barnes

Art Director
CJae Froshay

Imaging
Alfred Lau
Rosa C. See

Publisher
Mary D. Smith, M.S. Ed.

Author

Bev Dunbar

(Revised and rewritten by Teacher Created Resources, Inc.)

This edition published by ***Teacher Created Resources, Inc.***
6421 Industry Way
Westminster, CA 92683
www.teachercreated.com

ISBN-1-4206-3534-4

Table of Contents

Introduction

Measurement is an important aspect of any mathematics program. It is a practical way to apply number skills to solve problems specific to each student's stage of development. Both the classroom and the playground can be used when discussing measurement, and a wide variety of everyday objects can be used as measuring objects.

Math in Action: Time includes many action-packed ideas for developing skills in exploring, comparing, and using informal time concepts in fun, practical ways. The activities range from simple to super-challenging, to help support different ability groups.

Making your teaching life easier is a major aim of this series. The book is divided into sequenced units which contain activity cards and worksheets for small groups or a whole class to explore. You will also find easy-to-follow instructions, with assessment help in the form of clearly stated skills linked to a record sheet (page 93).

Each activity is designed to maximize the way in which your students construct their own understanding about time. The activities are generally open-ended and encourage each student to think and work mathematically. The emphasis is always on practical manipulation of materials and the development of language and recording skills.

Have fun exploring time concepts with your students.

How to Use This Book

❑ Teaching Ideas

Included in this book are many exciting teaching ideas which have been divided into five sections to assist your lesson planning for the whole class or small groups. Each activity has clear learning outcomes and easy-to-follow instructions. Activities are open-ended and encourage your students to think for themselves.

❑ Reproducible Pages

In this book are many reproducible pages. Below are some examples of the different types of pages which are included in this book.

Reusable Worksheets

(e.g., page 69, How Far Can I Go in a Minute?)
These worksheets support free exploration, as well as structured activities. They are great for reuse with small groups.

Discussion Cards

(e.g., page 57, Which Takes Longer?)
Cut these out, shuffle, and use over and over again for small-group games. Copy each set in a different color for easy classroom managment.

Activity Cards

(e.g., page 36, Ways to Work with Weeks)
Use these cards as an additional stimulus in group work. The language is simple and easy-to-follow. Encourage your students to invent their own activity cards too. You can laminate them so that they last for years.

❑ Skills Record Sheet

The complete list of learning outcomes is available on page 93. Use this sheet to record individual student progress.

❑ Sample Weekly Lesson Plans

On pages 94 and 95, you will find examples of how to organize a selection of activities for Exploring Days of the Week and Telling Time in Hours as five-day units for your class. A blank weekly lesson plan is included on page 96 for your individual use.

Exploring Time Language

In this unit, your students will do the following:

- ❏ Identify differences between daytime and nighttime
- ❏ Use everyday language related to day and night
- ❏ Identify and sequence events within one day

(The skills in this section are listed on the Skills Record Sheet on page 93.)

Day Detectives

Skills

- ❑ Identify differences between daytime and nighttime
- ❑ Use everyday language related to day and night

Grouping

❑ whole class ❑ individuals

Materials

- ❑ Day Detectives worksheet (page 7)
- ❑ posters, magazines, scissors
- ❑ picture books about day or night
- ❑ job discussion cards (pages 8 and 9)

Directions

- ❑ Have students participate in the following activities both inside your classroom and then outside on the playground. Say the following to students:

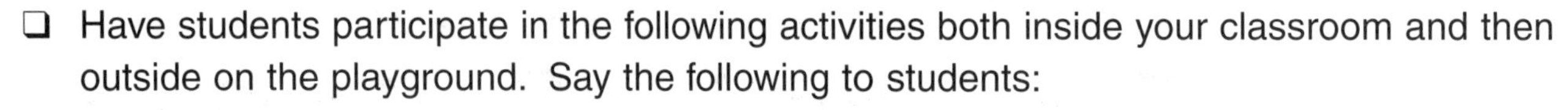

  "Look around you. How can you tell whether it is daytime or nighttime?"

  "Close your eyes. What clues tell you whether it is daytime or nighttime?"

  "Block your ears. What evidence is there that it is daytime or nighttime?"
- ❑ Discuss their discoveries together. Ask, "Are the clues the same? What types of things do you do in the daytime? Why? What types of things do you do at night? Why?"
- ❑ Record some of their discoveries on the worksheet.

Variations

- ❑ Ask students, "What types of jobs do people do in the daytime? At night? Sort the job cards into three piles—day jobs, night jobs, and day or night jobs. Add more cards of your own, including jobs your parents do."
- ❑ Discuss clues on posters showing day and night themes.
- ❑ Cut out, sort, and glue magazine pictures into day and night sets.
- ❑ Read stories related to day or night themes. (e.g., Berenstein S. J. (1971) *Bears in the Night*. Collins and Harvill.) Look for clues that tell you the time of day.

Day

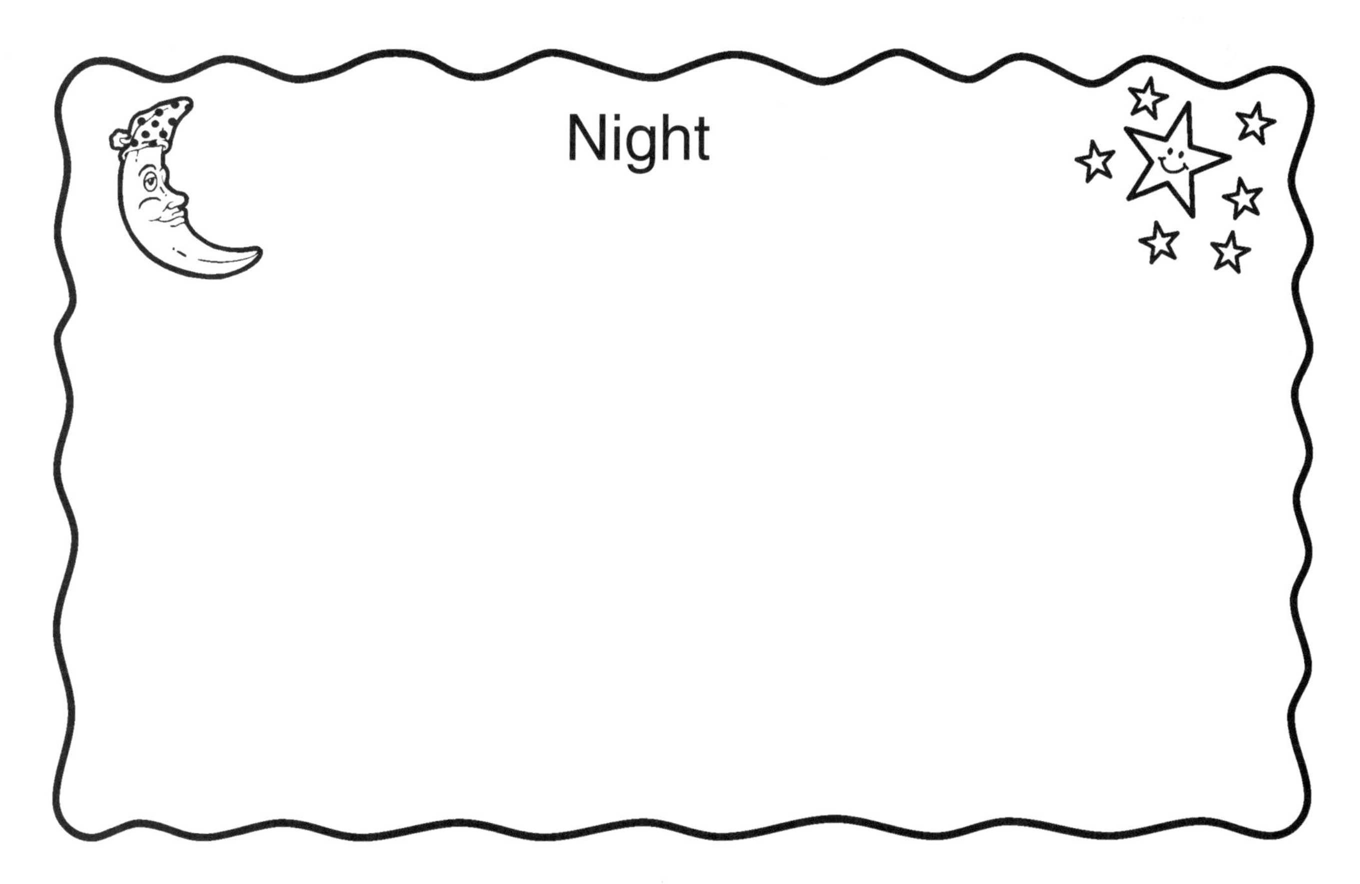
Night

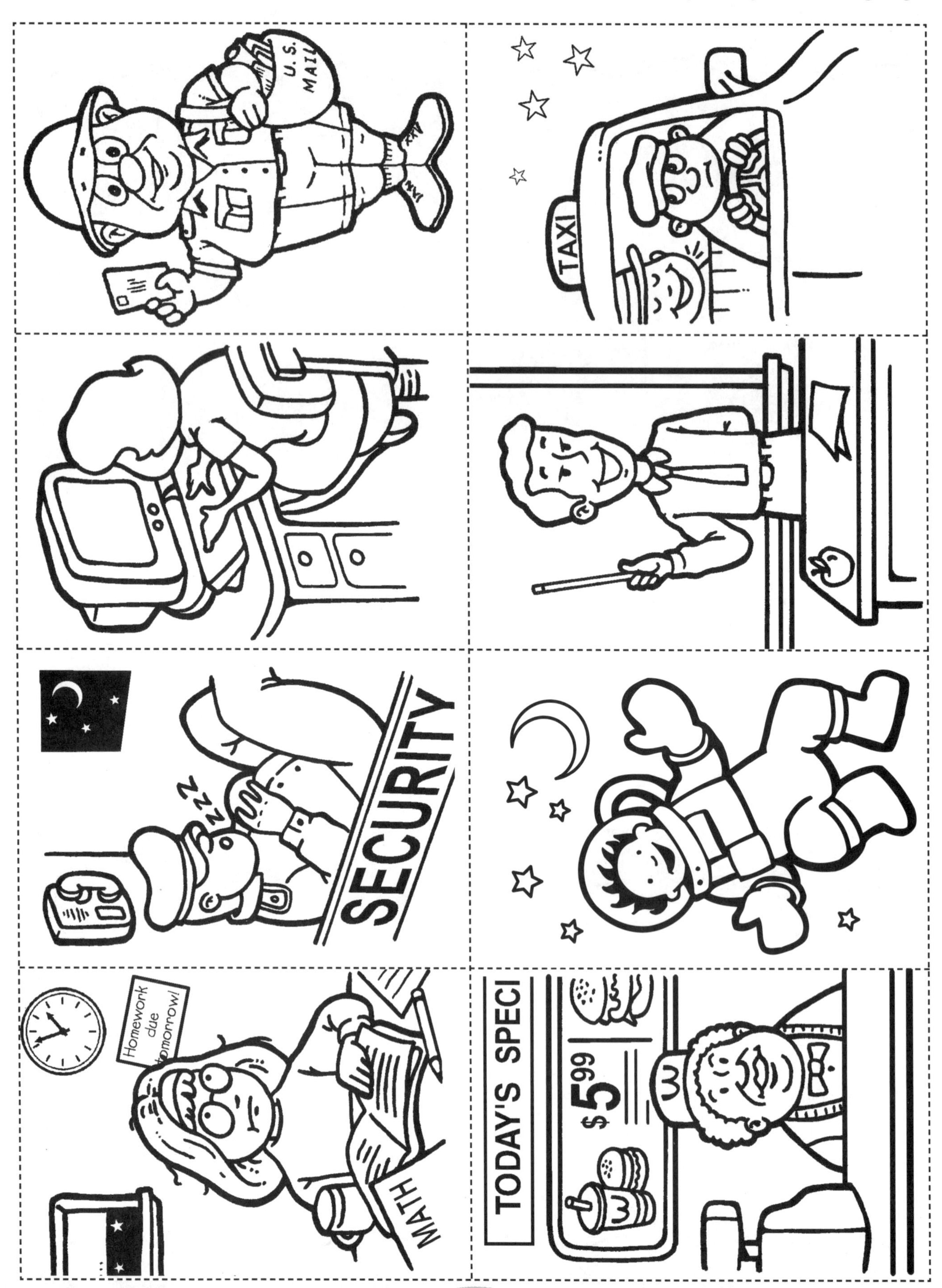
U.S. MAIL
TAXI
SECURITY
Homework due tomorrow!
MATH
TODAY'S SPECI
$5 99

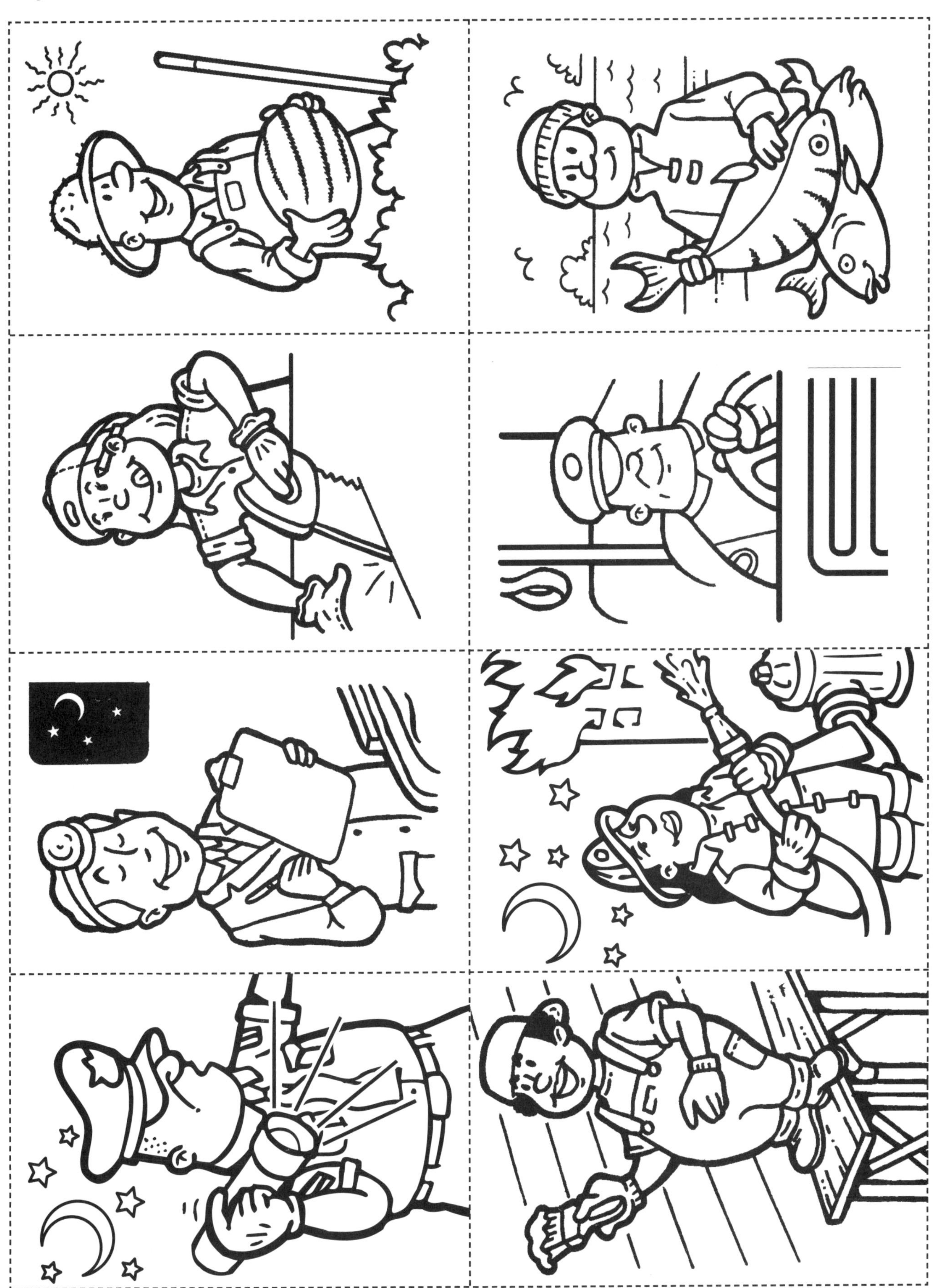

What's in a Day?

Skills

- ❑ Identify differences between daytime and nighttime
- ❑ Use everyday language related to day and night
- ❑ Identify and sequence events within one day

Grouping

- ❑ whole class
- ❑ small groups

Materials

- ❑ What's in a Day? picture sorting sets (pages 11–13)
- ❑ What's in a Day? discussion cards (pages 14 and 15)

Directions

- ❑ Ask students, "How do you know when daytime begins?" (e.g., When the sun appears to rise over the horizon.") Discuss different words to describe the start of a day. (e.g., *sunrise, dawn*)
- ❑ Ask students, "How do you know when nighttime begins?" (e.g., When the sun appears to set below the horizon.") Discuss different words to describe the end of a day. (e.g., *sunset, twilight, evening*)
- ❑ Ask students, "What's the first thing you do when you get up in the morning? What's the last thing you do before you get into bed at night?"
- ❑ Ask students, "What words do you use to describe time passing in a day?" (e.g., "What do you call the time before lunch? After lunch? What do you call the middle of the day?")
- ❑ Break students into groups with a set of sorting cards for each group. Discuss the events occurring in each picture. Sort them into a sequence from what happens first to what happens last. Answers may differ with each group.

Variations

- ❑ Model a daily cycle of events on a circular chart with drawings or picture cutouts. Discuss *before* and *after*.
- ❑ Have students shuffle the discussion cards. Have them turn over any three cards and sort them into order based on what happens first. Ask, "Which one comes first in the day? Which one comes next?"

Picture Sorting Set 1

Picture Sorting Set 2

Picture Sorting Set 3

Discussion Cards

Discussion Cards

Food Times

Skills

- ❑ Identify differences between daytime and nighttime
- ❑ Use everyday language related to day and night
- ❑ Identify and sequence events within one day

Grouping

❑ whole class ❑ pairs

Materials

- ❑ a few ingredients from a typical breakfast, lunch, and dinner
- ❑ Food Times pictures (pages 17 and 18)
- ❑ magazine pictures, scissors, glue

Directions

- ❑ Discuss names for food times during a day. (e.g., breakfast, morning snack, lunch, afternoon snack, dinner)
- ❑ List foods students might eat or drink at each time. Ask them, "In which ways are these meals similar? In which ways are they different?"
- ❑ Hold up different meal ingredients one at a time. Ask students, "When would you eat this—in the morning, middle of the day, or for dinner at night? Why?"
- ❑ Break students into pairs. Have them cut out the Food Times pictures. With their partners, have students discuss the sort of food shown on each card. Have them sort the cards into groups. (e.g., These foods can be eaten at any time.)
- ❑ Have students try to sort all the cards into time order from the start of the day to the finish. Have each group compare its order with another group. Ask, "Are some foods always sorted the same way?"

Variations

- ❑ Have students look for food pictures to cut out, sort, and glue into a workbook showing food for different times of the day.
- ❑ Have students investigate different breakfast menus from around the world. (e.g., What would you eat for breakfast if you grew up in India?)

OATMEAL
JAM

What's My Story?

Skills

- ❑ Identify and sequence events within one day

Grouping

- ❑ whole class ❑ individuals

Materials

- ❑ a favorite picture book story
- ❑ What's My Story? strips (page 20 or 21–22)
- ❑ What's My Story? comic strip cards (page 23)
- ❑ paper, scissors, glue

Directions

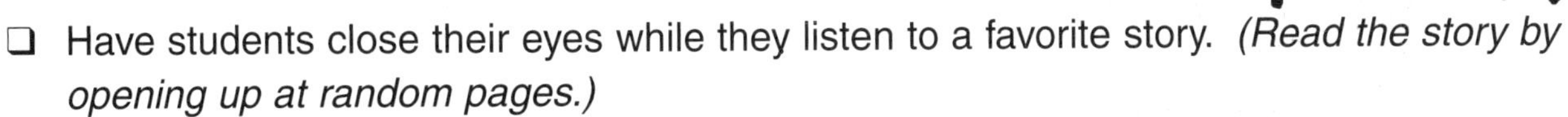

- ❑ Have students close their eyes while they listen to a favorite story. *(Read the story by opening up at random pages.)*
- ❑ Ask students, "What do you notice?" (e.g., The story was all mixed up. It did not start from the beginning.)
- ❑ Discuss the fact that most stories, rhymes, songs, or comic strips have a recognizable beginning, middle, and end.
- ❑ Have students sit in a circle. Agree on a favorite story to tell together. The first person starts the story anywhere except at the normal beginning. Then he or she selects someone to be the second storyteller. Continue until the complete story has been told, but all mixed up.
- ❑ Read the sentence strips together. Have students cut out and glue the strips into a correct story sequence on their papers. Have them draw a picture to match. (*Note:* The answers for the story on page 21 are located on page 22.)

Variations

- ❑ Class Challenge: Have each student tell his or her own mixed-up story or rhyme and then ask a friend to say which part of their story should come first.
- ❑ Have students cut out a favorite comic strip (or use the comic strip cards on page 23), mix them up, and then sort the strip into correct time order. Ask, "Which event happens first?"
- ❑ Have students cut up lines to a favorite song. Have them sort the song into correct time order. Ask them, "Which line comes first?"
- ❑ Have students create their own story sentences to sequence into strips.

She sat on baby bear's chair, and it broke.

The three bears went for a walk in the woods
while their porridge cooled.

The bears never saw Goldilocks again.

Goldilocks ran away when the bears
came back and frightened her.

Soon after, Goldilocks discovered the bears'
house and went inside.

She went upstairs. She was so tired
that she fell asleep in baby bear's bed.

She tasted baby bear's porridge.
It was so delicious that she ate it all up.

Then he disappeared into the clouds!

When I went in to look, I couldn't believe it!
My brother had turned into an alien!

Suddenly, the ceiling opened up, and
he floated up into the sky.

His ears grew into points, his legs were long
and skinny, and he wore a space suit.

Now what am I going to tell
Mom and Dad?

I heard a strange sound
in my brother's room.

I waved goodbye, and he
sent me a ray of silver light!

Answer Key for Page 21

I heard a strange sound
in my brother's room.

When I went in to look, I couldn't believe it!
My brother had turned into an alien!

His ears grew into points, his legs were long
and skinny, and he wore a space suit.

Suddenly, the ceiling opened up, and
he floated up into the sky.

I waved goodbye, and he
sent me a ray of silver light!

Then he disappeared into the clouds!

Now what am I going to tell
Mom and Dad?

FIDO

Exploring Days of the Week

In this unit, your students will do the following:

- ❏ Name and order the days of the week
- ❏ Identify and sequence events within a week
- ❏ Use the terms *yesterday, today,* and *tomorrow*

(The skills in this section are listed on the Skills Record Sheet on page 93.)

Turn-a-Week

Skills

- ❑ Name and order the days of the week

Grouping

- ❑ whole class ❑ individuals

Materials

- ❑ Turn-a-Week Wheel (page 26)
- ❑ The Days of the Week (page 27)
- ❑ scissors, colored pencils, brads

Directions

- ❑ Make a sample Turn-a-Week Wheel. (See bottom illustration.)
- ❑ Tell the story of how people long ago discovered seven wandering bright objects in the night sky. These were the sun and the moon and five planets: Venus, Jupiter, Saturn, Mars, and Mercury. They thought these objects were messengers of the gods who controlled the seven days of the week. Tell students that in countries all around the world, people made up names for the days of the week often based on these wanderers and their gods. Tell them our Sunday and Monday come from the old names Sun day and Moon day. Tiw was the old Scandinavian god of law associated with Mars. Tiw's day became our Tuesday. Woden was the old Scandinavian king of the gods. Woden's day became our Wednesday. Thor was the god of thunder. Thor's day became our Thursday. Freya was the old Scandinavian god of peace. Freya's day became our Friday. Saturn was the old Roman god of farming. Saturn's day became our Saturday.
- ❑ Tell students the days of the week are a cycle, like a wheel turning around and around in exactly the same order each time. Discuss the sample Turn-a-Week. Show how the pointer moves around clockwise from day to day.
- ❑ Demonstrate how to cut out and color the wheel and the pointer. Explain how to attach a brad to join them in the center. The pointer goes at the back. Have students use this to practice saying the days of the week starting at any day.

Variations

- ❑ Read *The Very Hungry Caterpillar* by Eric Carle (London: Puffin, 1970).
- ❑ Explore versions of *Snow White and the Seven Dwarfs*.
- ❑ Have students investigate and record names for days of the week in other languages. Have students use page 27 as a start.

Today is
Monday
Tuesday
Wednesday
Thursday
Friday
Saturday
Sunday

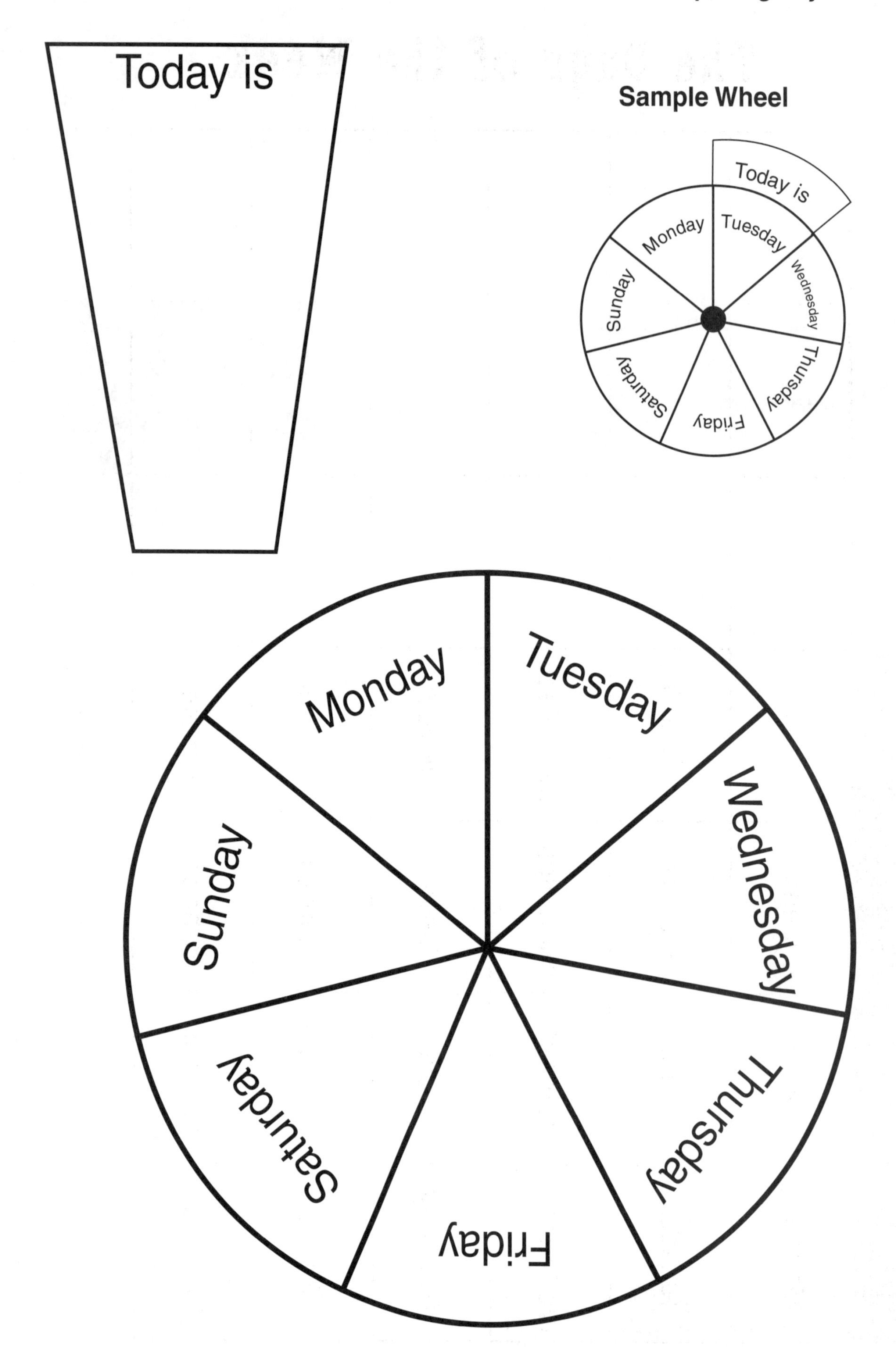
Today is
Sample Wheel
Today is
Monday
Tuesday
Wednesday
Thursday
Friday
Saturday
Sunday
Monday
Tuesday
Wednesday
Thursday
Friday
Saturday
Sunday

The Days of the Week

English	French	Spanish	Italian			
Monday	lundi	lunes	lunedi			
Tuesday	mardi	martes	martedi			
Wednesday	mercredi	miércoles	mercoledi			
Thursday	juedi	jueves	giovedi			
Friday	vendredi	viernes	venerdi			
Saturday	samedi	sábado	sabato			
Sunday	dimanche	domingo	domenica			

What's in a Week?

Skills

- ❑ Identify and sequence events within a week
- ❑ Use the terms *yesterday, today,* and *tomorrow*

Grouping

❑ whole class ❑ small groups

Materials

- ❑ magazines, scissors, glue
- ❑ days of the week weather chart (page 30)
- ❑ paper, pencils
- ❑ days of the week worksheet (page 31)
- ❑ Days-of-the-Week Cards (page 29)

Monday

Directions

- ❑ Ask students, "What day is it today? What's special about it? What day was it yesterday? What day will it be tomorrow?"
- ❑ Discuss the term *weekend.* Ask students, "Why don't you go to school for all seven days each week?"
- ❑ Form three groups with a set of cards each.

  Group A: Identify and discuss weekly school routines. (e.g., assembly on Mondays)

  Group B: In your family, which events occur on the same day each week? (e.g., Visit Grandma on Friday night)

  Group C: What favorite TV programs occur each day?
- ❑ Have students draw and cut out pictures or write about their special events for each day. Record these on a large class display labelled with the seven days of the week.

Variations

- ❑ Have students cut out and glue the seven days of the week in order onto their papers. Have them draw something special that happens on each day.
- ❑ Have students keep a daily record of the weather. Have them count how many cold, warm, and hot days in a week. Use the form on page 30.
- ❑ Have students write a daily diary for one week, recording what is special for them about each day. Use the log on page 31.

Days-of-the-Week Cards

Directions: Write or draw a description of the weather for each day of the week.

Weather Chart	Monday	Tuesday	Wednesday
Thursday	Friday	Saturday	Sunday

Directions: Write or draw the special things that happened on each day of the week.

Monday	Tuesday

Wednesday	Thursday

Friday	Saturday/Sunday

Ways to Work with Weeks

Skills

- ❑ Name and order the days of the week
- ❑ Identify and sequence events within a week
- ❑ Use the terms *yesterday, today,* and *tomorrow*

Grouping

- ❑ small groups

Materials

- ❑ Days-of-the-Week Spinners (page 33), pencils, paper clips
- ❑ Days-of-the-Week Cards (page 29)
- ❑ Position Cards (pages 34 and 35)
- ❑ activity cards (pages 36–37)
- ❑ craft sticks, felt pens

Directions

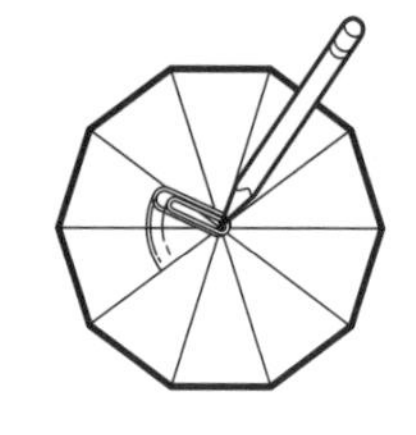

- ❑ Reproduce the spinners on heavy cardstock paper. Follow the directions on page 33.
- ❑ Copy and cut out the seven position cards.
- ❑ Make a set of days-of-the-week sticks by writing each day in the center of a craft stick with a felt pen.
- ❑ Class Challenge A: Form four groups. Briefly explain the Spin-a-Week activities (page 36). Ask if students have any questions. Give each group a spinner and an activity card. Group 2 will need a set of Position Cards, shuffled, and placed face down. Group 4 will need a set of days-of-the-week sticks.
- ❑ After a suitable time (e.g., five minutes), have students add up their groups' scores and rotate activities. At the end of this session, find out which group has scored the most points.

Variations

- ❑ Class Challenge B: Form four groups. Briefly explain the Deal-a-Week activities (page 37). Give each group (except Group 1) a set of shuffled Days-of-the-Week cards (page 29) face down and an activity card. Group 1 will need a set of shuffled Position Cards. Group 2 will need a set of Days-of-the-Week Cards for each student. Ask if students have any questions.
- ❑ Have students play the game on page 38 with a partner. Each group will need a set of Position Cards and a Days-of-the-Week spinner.

Days-of-the-Week Spinner

Color and cut out the spinner. Put a paper clip over the tip of a pencil. Place the tip of the pencil on the center of the spinner and spin the paper clip.

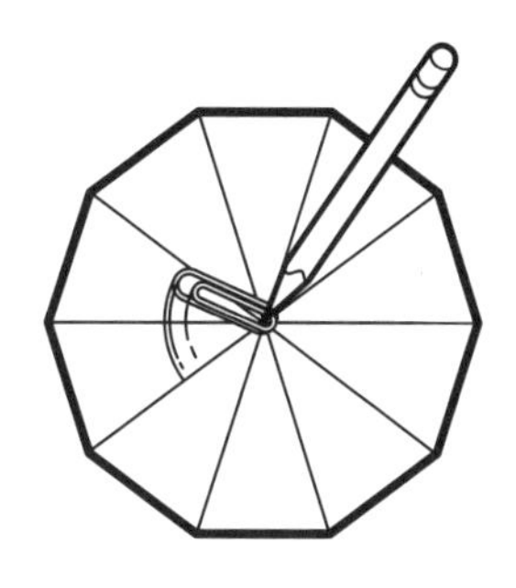

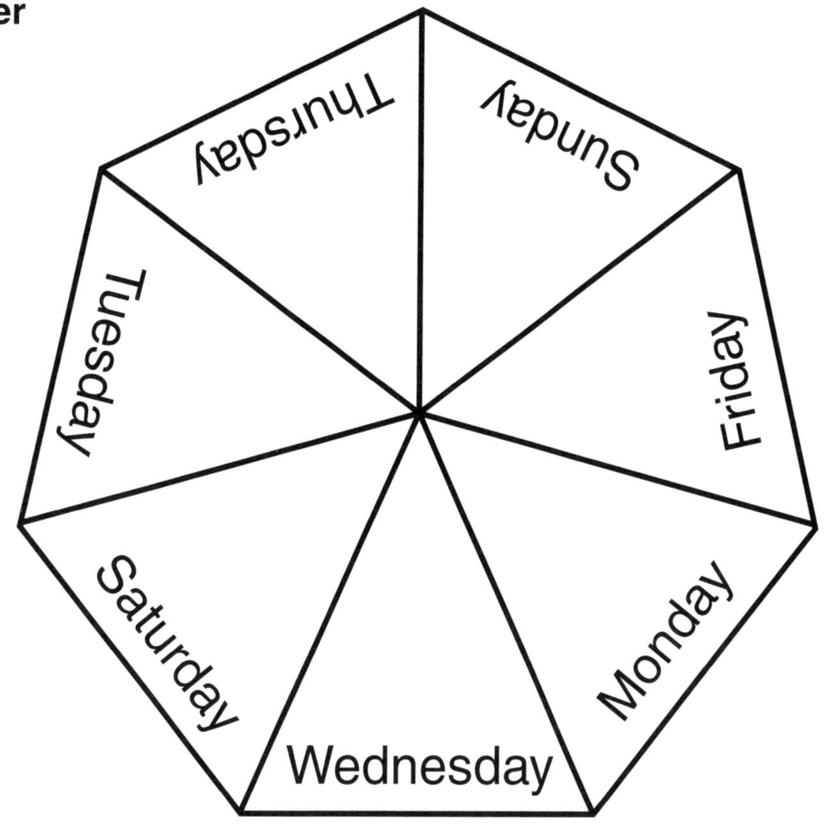

Days-of-the-Week Spinner

Color and cut out the spinner. Put a paper clip over the tip of a pencil. Place the tip of the pencil on the center of the spinner and spin the paper clip.

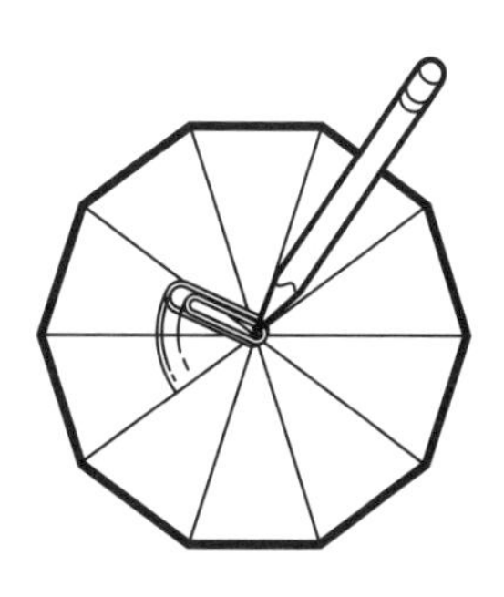

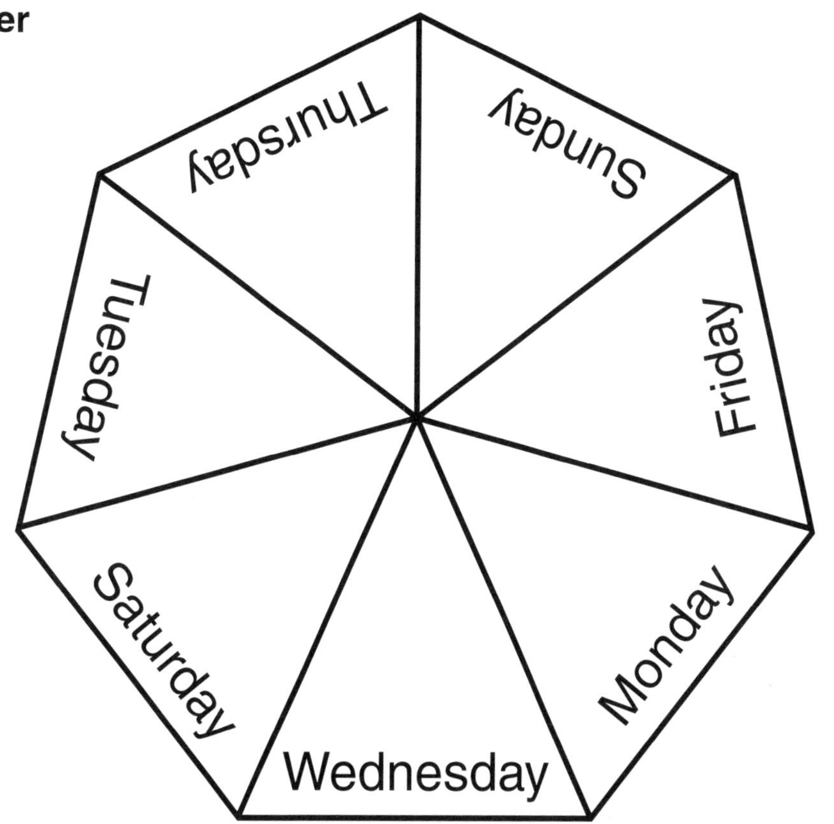

Position Cards

the day after

yesterday

tomorrow

the day before

today

the day after tomorrow

the day before yesterday

Position Cards

last night

tonight

this morning

this afternoon

tomorrow night

tomorrow afternoon

yesterday morning

Spin-a-Week 1

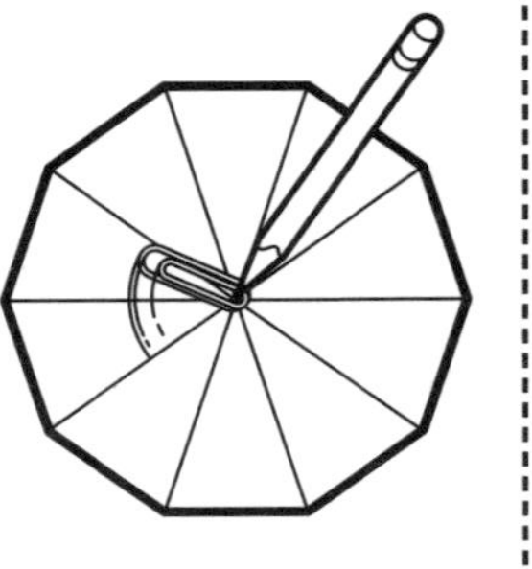

Spin the spinner.

Starting from this day, say the days of the week in order to score two points.

Score two extra points if you can name them backwards, too.

Spin-a-Week 2

Spin the spinner.

Turn over a position card. Score two points if you can name the matching day.

Example: tomorrow Friday

Spin-a-Week 3

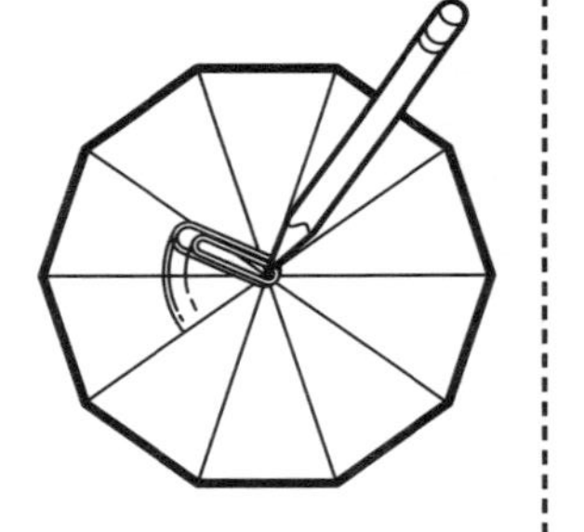

Spin the spinner.

Score two points for each different event you can name that happens on this day.

Score 0 points if someone has already said your idea.

Spin-a-Week 4

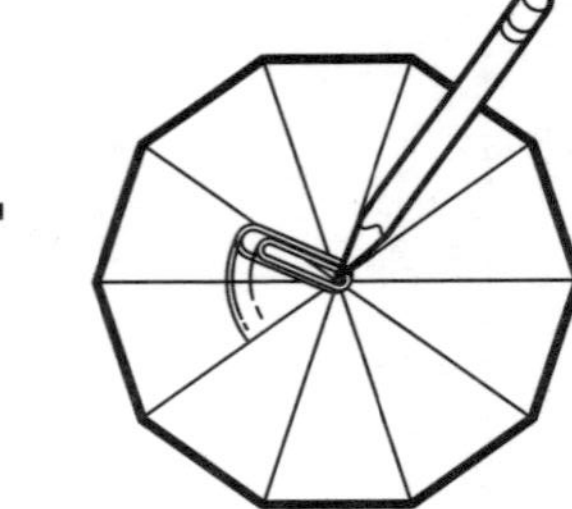

Spin the spinner. Take a matching craft stick.

Score 5 points when you collect enough sticks to make 3 days in a row.

Deal-a-Week 1

Turn over a position card.

Score five points if you can name the matching day.

Example: | the day before yesterday |

Deal-a-Week 2

Each of you takes a set of days-of-the-week cards and mix them up face down. Race to turn them over and sort them into order. Time each person.

The fastest person scores five points.

Deal-a-Week 3

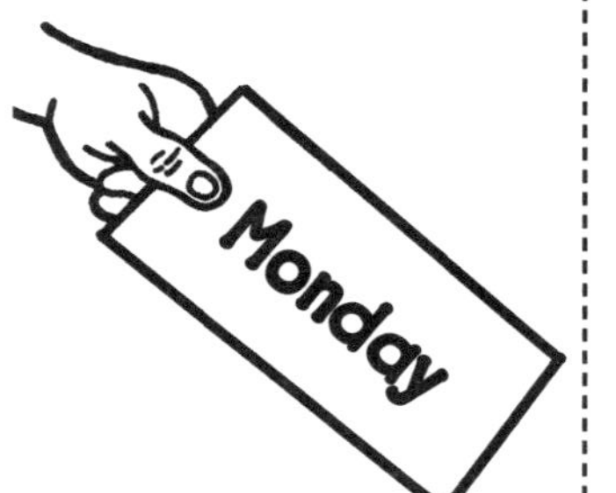

Work with a partner. Each of you takes a day-of-the-week card. Score five points if you can say the days of the week, starting at your card and finishing at your partner's card.

Deal-a-Week 4

Take a day-of-the-week card.

Score five points if you can count how many days from today until the day you selected.

Directions: Make a pile of the position cards, face down. With a partner, take turns spinning the Days-of-the-Week spinner. When the spinner stops, look at the day of the week. Turn over a position card. If it is a match, take a point card (shown below) and spin again. If it is not a match, your turn is over. When the game is over, add up your point cards to see who is the winner. (Additional information on making matches: If, for example, you are playing the game on Friday, and the spinner comes up as "Thursday," and the position card is "yesterday," then you have a match!)

Exploring Months of the Year

In this unit, your students will do the following:

- ❑ Name and order the months of the year
- ❑ Identify and sequence special events within a year
- ❑ Name and order the seasons
- ❑ Recognize differences between months
- ❑ Use a calendar to show a day and date

(The skills in this section are listed on the Skills Record Sheet on page 93.)

What's in a Year?

Skills

- ❑ Name and order the months of the year
- ❑ Identify and sequence special events within a year

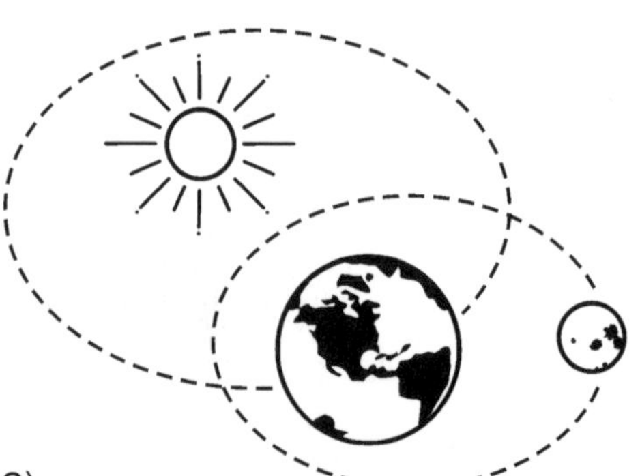

Grouping

❑ whole class ❑ small groups

Materials

- ❑ months cards (page 41)
- ❑ blank calendar (page 42)
- ❑ chart displaying months in different languages (page 43)
- ❑ chart paper, scissors, glue

Directions

- ❑ Have students look around them. Ask them, "Can you tell what month it is?" Discuss.
- ❑ Recite the months of the year together. Have students do this as fast as they can.
- ❑ Tell students that people long ago watched the sky at night and noticed that the moon's cycle is fairly regular. Tell them our months are based on these earlier observations and that our word *month* comes from the word *moon*. Tell them each month is about four weeks. Tell students that people also noticed that the earth goes around the sun in a regular cycle. Every 12 months a new cycle begins, and this cycle is called a *year*.
- ❑ Tell students that the months of the year are like a circle and that they follow each other in order. Have each student tell the person beside him or her the month in which he or she was born. Ask, "Do you know which month comes first?" Inform them that the months of the year were named long ago by the Romans. Some of the months were named after famous emperors. (e.g., July is named after Julius Caesar who lived more than 2,000 years ago.)
- ❑ Have students cut out one set of months cards and glue them in order from *January* to *December* to make a class chart or poster.
- ❑ Have students work in small groups and create a story or rhyme to help them remember the order of the months. (e.g., Think of the starting letter of each month—Jolly Farmer Mack Ate My Jam Jelly Apricot Sandwich On Nick's Desk.)
- ❑ Discuss their suggestions and select the most interesting ones to recite as a class.

Variations

- ❑ Make a reduced copy of the months cards. Have students cut them out, order them, and glue them into a workbook.
- ❑ Have students look up into the sky at night and keep a record of what the moon looks like, if they can see it. Have students use the blank calendar on page 42 to keep a record.
- ❑ Have students investigate words for the months in other languages. Have them use the cards on page 43 to investigate, arrange, and research. Also, students can play matching games with the two languages (French and Spanish).
- ❑ Have students investigate the origin of each month's name.

November	March
June	September
January	May
August	December
October	February
April	July

Directions: Use the calendar below to chart the moon's appearance all month long. Each night, take a look at the moon and draw what you see. It will only change a little bit each night, but through the entire month, it will change a lot. Make twelve copies of this page if you would like to draw the moon for an entire year!

The Months in Other Languages

English	French	Spanish
January	janvier	enero
February	fevrier	febrero
March	mars	marzo
April	avril	abril
May	mai	mayo
June	juin	junio
July	juillet	julio
August	aout	agosto
September	septembre	septiembre
October	octobre	octubre
November	novembre	noviembre
December	decembre	diciembre

What's in a Season?

Skills

- ❑ Identify and sequence special events within a year
- ❑ Name and order the seasons

Grouping

❑ whole class ❑ small groups ❑ individuals

Materials

- ❑ posters, pictures related to the four seasons
- ❑ recording of Vivaldi's *Four Seasons*
- ❑ What's in a Season? worksheets (pages 45 and 46)

Directions

- ❑ Tell students to look around them. Ask them, "Can you tell what season it is?" Discuss.
- ❑ Tell students the following: "In some countries, the cycle of the months through the year can be broken into four distinct groups—spring, summer, autumn, and winter. In some parts of the world, the seasons are not so distinct."
- ❑ Discuss the months of the year matched to each season.
- ❑ Tell students the following: "Vivaldi was an Italian composer who lived in Italy about 300 years ago. In Italy you can see the seasons easily." Play extracts from one of Vivaldi's seasons. Ask students what images the music creates. Ask, "Can you guess which season he was picturing?"
- ❑ Discuss the special features of each season where students live. Form small groups. Discuss features of the local weather, changes in the environment around them, the clothes they wear, the food they might eat or the sports they might play in each season.
- ❑ Finish by discussing the four children on page 45. Tell your students to imagine each one in a different setting. Ask students, "What extra features can you add to show the children in four different seasons?" (e.g., clothes—a hat and swim suit, weather—trees with leaves falling down, blowing in the wind) Next, tell students to imagine the scene on page 46 in four different seasons and add details.

Variations

- ❑ Have students explore features in famous paintings that can be related to one of the four seasons.
- ❑ Have students make a birthday graph. (e.g., use the months cards from page 41) Ask them, "Which season is the most popular for birthdays? Which season is the least popular?"
- ❑ Have students investigate names for the seasons in other languages.

Toss-a-Month

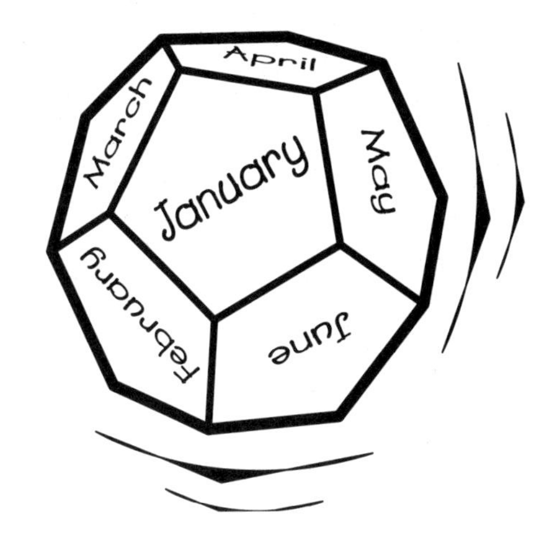

Skills

- ❑ Name and order the months of the year
- ❑ Name and order the seasons

Grouping

- ❑ whole class
- ❑ pairs

Materials

- ❑ months of the year poster
- ❑ Toss-a-Month dodecahedron die (page 48)
- ❑ one sample large dodecahedral die
- ❑ Toss-a-Month Discussion Cards (page 49)
- ❑ Toss-a-Month Birthdays (page 50)
- ❑ glue
- ❑ scissors

Directions

- ❑ Have students tell you all they know about the months of the year. (e.g., January is the first month. September is the first month of autumn.)
- ❑ Have students toss the months die. Ask them questions about the month the die shows. (e.g., "What is special about this month? In which season is this month? Which month comes just before it? Just after it? Who has a birthday in this month?") Have students use the months poster as a reference, if necessary.
- ❑ Have students make their own die by coloring and then cutting out the die, including the tabs. Have them fold along all the dashed lines and glue the tabs under to create a ball-like shape.
- ❑ Have students use the die in partner games, making up their own time questions. Or, have them use the discussion cards. Have students shuffle the cards, then place them face down. Have students toss the die, turn over the top card, and say the matching month to their partners.

Variation

- ❑ Fill in the chart on page 50 to remember student birthdays.

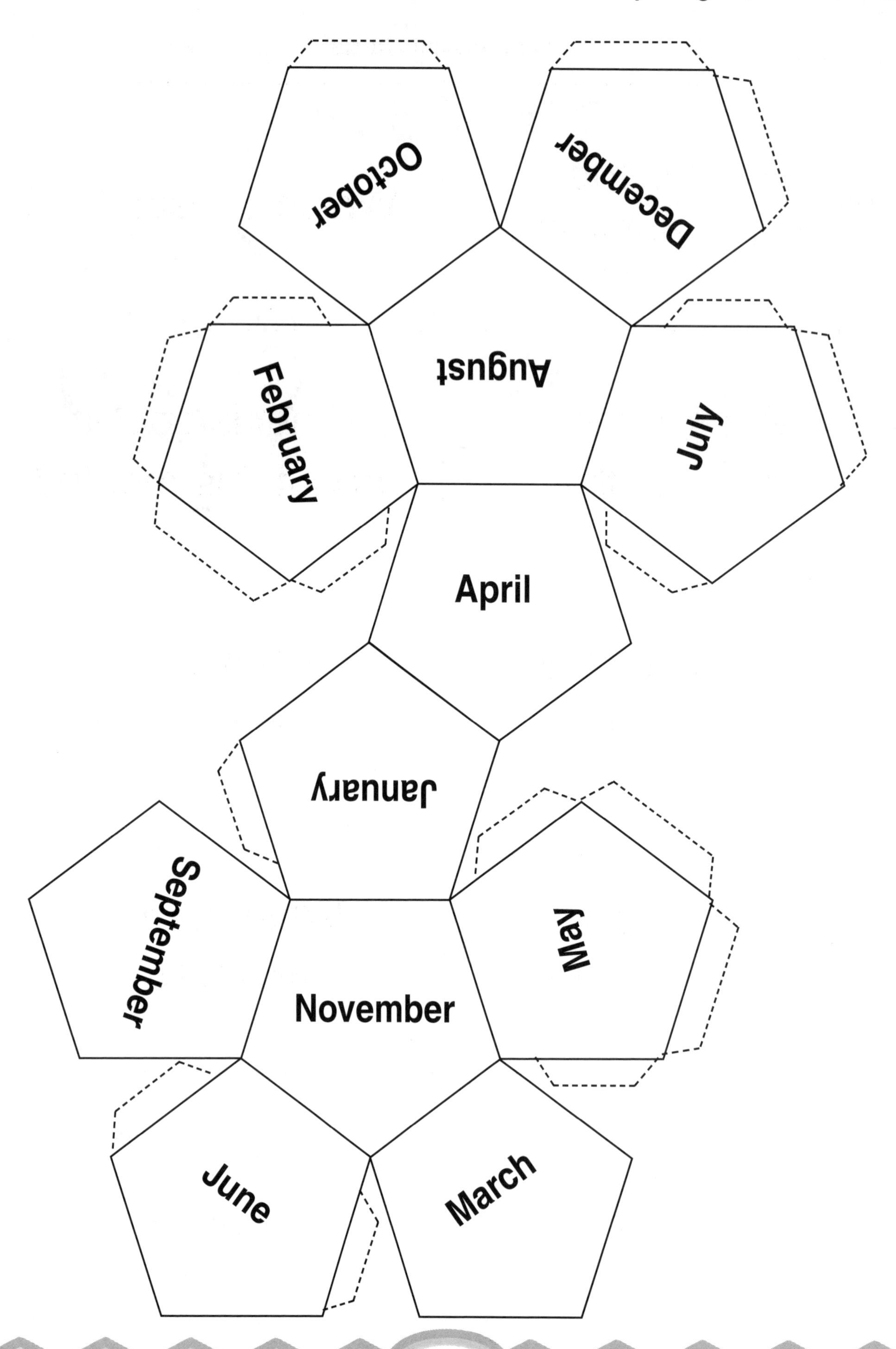
October
December
August
February
July
April
January
September
November
May
June
March

Discussion Cards

What is the month before?	Which season is it?
What is the next month?	What is special about this month?
Is it in the first half of the year?	What is two months before?
What is two months after?	What number month is it?
Is it the last month of a season?	Which months is it between?

Birthdays

NAME OF STUDENT / MONTHS	January	February	March	April	May	June	July	August	September	October	November	December

What's in a Date?

Skills

- ❑ Recognize differences between months
- ❑ Use a calendar to show a day and date

Grouping

- ❑ whole class

Materials

- ❑ examples of different calendars
- ❑ How Many Days? rhyme (page 52)
- ❑ calendar example (page 53)
- ❑ blank calendar (page 54)

Directions

- ❑ Ask students, "What is a calendar? For what do you use it? Why did people invent it?" Discuss examples of calendars together.
- ❑ Tell students that people all over the world have agreed on a common way to refer to time. Many cultures have their own time recording system but still recognize the same calendar system we use. This system is now over four hundred years old.
- ❑ Tell students that each year is divided into 12 months. The number of days in each month is not always the same. They were originally related to the phases of the moon in the sky. The Romans decided to make some months shorter and some months longer. (e.g., Emperor Augustus wanted his month to be as long as Julius Caesar's month. July had 31 days, so he declared that August would have 31 days, too.) Ask students how many days are in all the other months.
- ❑ Tell students the way Earth travels around the sun is not as regular as people would like it to be. So in the past, the written number for the year sometimes became out of step with the actual season. To help keep our years regular, it was decided to add an extra day every four years. This is called a Leap Year. This extra day is added to February because it has the smallest number of days.
- ❑ Tell students each day of the year has its own special number. When they write the date, there are many different ways to do it. They can write it as words with the year. (e.g., the second of October, 2001) They can write it as numbers and words. (e.g., 30th of June, 2003) They can write it all as numbers. (e.g., 2-25-02 or 4/13/04) Ask students, "Which way do you like best?" Practice writing the date together.

Variations

- ❑ Have students learn a rhyme to memorize the number of days in a month. (e.g., "How Many Days?" Or, have them create a rhyme of their own.)
- ❑ Have students create their own calendars. There is a blank template on page 54 and a sample calendar on page 53.

How Many Days?

April, **June**,
And **September**

Thirty days,
Like **November**.

February
Very fine,

Twenty-eight,
Or twenty-nine

All the rest
Have thirty-one.

Name them all,
Then it's done.

January, **March**,
May, **July**, **August**,
October, **December**!

JULY

Sunday	Monday	Tuesday	Wednesday	Thursday	Friday	Saturday
				1 Library Piano Practice	2 Soccer	3 Swim Lessons
4 Fourth of July	5 Piano Practice	6 Soccer	7	8 I go to the dentist.	9 Dad's Birthday	10 Swim Lessons
11 Vacation Starts	12 Yeah!	13 Yeah!	14 Yeah!	15 Yeah!	16 Yeah!	17 Vacation Ends
18 Visit Grandma	19 Piano Practice	20 I get a tetanus shot.	21 Soccer	22 Library books are due.	23 Aunt's Wedding	24 Swim Lessons
25	26 Piano Recital	27	28 Soccer	29	30 Brother's Birthday	31 Scout Campout

Exploring Informal Time

In this unit, your students will do the following:

- ❏ Compare the duration of two or more events
- ❏ Use informal time units to show time passing
- ❏ Demonstrate an awareness of seconds
- ❏ Demonstrate an awareness of minutes

(The skills in this section are listed on the Skills Record Sheet on page 93.)

Which Takes Longer?

Skill

- ❑ Compare the duration of two or more events

Grouping

- ❑ whole class
- ❑ small groups

Materials

- ❑ Which Takes Longer? discussion cards (pages 57–59)

Directions

- ❑ Ask students, "What are some things that take a short time to complete?" (e.g., blink an eye; stand up; drop a coin to the floor)
- ❑ Ask students, "What are some things that take a long time to complete?" (e.g., drive to visit Grandpa; grow up to be an adult)
- ❑ Ask students, "When you compare the time of two events, how do you know which one took longer?" (e.g., If they both start at the same time, the one that finishes first took the shorter time. It is faster. The one which took the longer time is slower.)
- ❑ Tell students that most races depend on time. The person who finishes first is usually the winner. Ask them, "What sort of time races can you invent?"
- ❑ Call out two events. (e.g., feed the cat and drink a glass of milk) Challenge the class to say which one will take the longer time to complete. Have them look for alternative explanations. (e.g., Does your cat expect you to stay with it while it eats? Is the glass large or small? Is the person drinking the milk a baby or a child?)
- ❑ Have students work in small groups with a set of shuffled cards face down. Have them select any two cards. Ask, "Which event would take the longer time? Why do you think this?"

Variations

- ❑ Have students create their own set of discussion cards.
- ❑ Have students select three or more cards and place the events in order from the shortest to the longest time to complete.

walk to school	get dressed
sing a song	put on my shoes
plant a seed and watch it grow into a flower	watch my favorite TV program
play a game	walk to the principal's office and back

write three lines of a story	eat an apple
take a bath	eat a candy bar
get a drink of water	sharpen my pencil
eat my lunch	write my name

fix a flat on my bike	make a sandwich
comb my hair	write a letter
wash the dishes	mail a postcard
make my bed	draw a picture

Just a Second

Skills

- ❑ Use informal time units to show time passing
- ❑ Demonstrate an awareness of seconds

Grouping

- ❑ whole class
- ❑ small groups

Materials

- ❑ stopwatch
- ❑ time logs (page 64)
- ❑ kitchen timers
- ❑ workbooks, pencils
- ❑ Timer Tasks activity cards (pages 61–63)

Directions

- ❑ Tell students that some things happen so quickly that they can blink and they are over. Have them name some things that happen this quickly. (e.g., a sneeze, a handclap, saying someone's name)
- ❑ Tell students that a very short time is called a *second*. Demonstrate with the stopwatch. Have students close their eyes and explore what 1–10 or more seconds feels like.
- ❑ Have students select a number of seconds. (e.g., five seconds) Have them predict and then check what they could do in this time. (e.g., snap your fingers five times)
- ❑ Explore what students can do in a very short time using different short time measures. Demonstrate how to use a kitchen timer. Set it for three seconds. Ask, "What can you do before the timer rings?" (e.g., touch your toes)
- ❑ Break into small groups with a different time challenge for each group. (e.g., five seconds, ten seconds, one minute) Have students invent their own timer challenges or use the sample cards provided.
- ❑ Have students record their favorite activities on page 64.

Variation

- ❑ Students can make their own timers using string and differently-sized stones. Tie the end of the string around a stone to make a pendulum. Hold the stone out with one hand while holding the other end of the string with the other hand. Let go of the stone so it will swing back and forth and see how long it takes to come to a complete stop. Experiment with longer and shorter string lengths and larger and smaller stones. Put a colored dot on each one. Make the blue timer last about five seconds, the red timer about 10 seconds, the yellow timer about 15 seconds, and the green timer about 20 seconds.

Timer Tasks

What can you do before one minute is up?

How many times can you write your name?

How many jumping jacks?

How many beads can you thread?

Record your results.

Timer Tasks

What can you do before two minutes are up?

Can you run to the fence and back?

Can you tie a set of shoelaces?

Can you read a page in a book?

Record your results.

Timer Tasks

What can you do before thirty seconds are up?

Can you touch your head, shoulders, knees, then toes?

How many times can you clap your hands?

Can you count backwards from 10 to 0?

Record your results.

Timer Tasks

What can you do before three minutes are up?

How many addition problems can you do?

How many blocks can you stack?

How many cats can you draw?

Record your results.

Timer Tasks

What can you do before forty-five seconds are up?

How many sit-ups?

How many times can you say your name?

Can you stand on one foot that long?

Record your results.

Timer Tasks

What can you do before fifteen seconds are up?

How many times can you jump?

How many times can you blink your eyes?

Can you write your name more than once?

Record your results.

Time Log

WHAT I DID	TIME								
	1 second	5 seconds	10 seconds	15 seconds	30 seconds	45 seconds	60 seconds/1 minute	2 minutes	3 minutes

Just a Minute

Skills

- ❑ Use informal time units to show time passing
- ❑ Demonstrate an awareness of minutes

Grouping

- ❑ whole class
- ❑ small groups

Materials

- ❑ stopwatch, one-minute egg timer, one-minute kitchen timer
- ❑ Just a Minute work cards (pages 66 and 67)
- ❑ Spot the Differences worksheets (page 68)
- ❑ various resources for making informal time measures (e.g., empty, plastic bottles with lids; sand; masking tape; scissors)
- ❑ access to Olympic records (e.g., *The Guiness Book of Records*, Dorling Kindersley (2000) The Olympic Games)
- ❑ How Far Can I Go In a Minute? worksheet (page 69)

Directions

- ❑ Ask students, "What is a minute? Why do you think people use minutes and not just seconds?" (e.g., Sometimes a second is too fast.) Tell them that a minute is a short time. It is 60 seconds long.
- ❑ Use the stopwatch. Have students close their eyes. Have them open their eyes when they think a minute has passed. Discuss their reactions together.
- ❑ Have students invent some minute challenges. (e.g., How many six-letter words can you write? What number can you count up to aloud?)
- ❑ Discuss other minute measures. Demonstrate how an egg timer and a kitchen wind-up timer work.
- ❑ Form small groups with a timer and work card for each group. Explain each task and collect any equipment they need. (e.g., The group making their own minute measure could construct their own sand timer using drink bottles.)

Variations

- ❑ Have students create their own Spot the Differences one minute challenge by reproducing a favorite cartoon or drawing their own pictures.
- ❑ Have students explore outdoor activities related to one minute. Use the How Far Can I Go In a Minute? worksheet to record some of their discoveries. Have them each invent their own activity to fill in at the bottom of the sheet. Students will need a measure device. (e.g., tape measure, yard stick)

How Long Is a Minute?

Imagine you had no clocks or watches.

How could you measure a minute exactly?

You can use a clock while preparing your time measure, but after that it should work on its own.

How Long Does It Take to Pass a Squeeze?

Sit in a circle.

Choose someone to go first.

The first person should squeeze the hand of the person next to him or her, then that person passes the squeeze to the next person and so on.

How long will it take to pass the squeeze all the way around the circle?

Guess the time first, then find a way to check your guess.

What Can an Olympic Champion Do in a Minute?

Investigate Olympic records.

Design a poster showing your discoveries.

Spot the Differences!

Look at the two pictures closely.

How many differences between the two pictures can you spot within one minute?

Spot the Differences!

Spot the Differences!

How Far Can I Go In a Minute?

	My Guess	Actual measure-ment
How far can I walk backwards?		
How far can I hop?		
How far can I walk sideways?		
How far can I walk three-legged with a partner?		
How far can I skip with a rope?		
How far can I continue jumping with both feet together?		
How far can I ____________?		

Telling Time in Hours

In this unit, your students will do the following:

- ❏ Demonstrate an awareness of an hour
- ❏ Tell time on the hour using an analog clock
- ❏ Tell time on the hour using a digital clock
- ❏ Tell time on the half-hour using an analog clock
- ❏ Tell time on the half-hour using a digital clock

(The skills in this section are listed on the Skills Record Sheet on page 93.)

How Long Is an Hour?

Skills

- ❑ Demonstrate an awareness of an hour

Grouping

- ❑ whole class ❑ small groups

Materials

- ❑ one-hour timers (e.g., clocks, stopwatches)
- ❑ an alarm clock
- ❑ paper strips, pencils
- ❑ Time Detective word puzzle (pages 72 and 73)
- ❑ Things That Take an Hour (page 74)

Directions

- ❑ Tell students that people long ago were not so concerned about very short periods of time. They divided their day into longer periods called *hours*. Ask students, "What do you know about an hour?" (e.g., It is 60 minutes long. There are 24 hours in a day. We go to school for six hours. Most people sleep for about eight hours every night.)
- ❑ Ask students, "How can you measure an hour?" (e.g., stopwatch, wrist watch, clock) Discuss suggestions for hour measures. (e.g., Set the alarm clock for one hour.)
- ❑ Ask students, "What do you know that lasts for exactly one hour?" (e.g., My favorite TV program; our math lesson on Friday.) Have students record as many one-hour events as they can on page 74.
- ❑ Form small groups with 12 paper strips for each group. Have students record four events that they think take less than one hour, four events that they think take about one hour, and four events that they think take more than one hour.
- ❑ Have students shuffle their strips and exchange them with another team. Ask them, "How long does it take you to sort them into the three groups?"
- ❑ Have students record some of their discussions as a class display.

Variations

- ❑ Have students look at the Time Detective word puzzle. There are 18 hidden words, some back-to-front, up, or down. Students try to find each of the words on the list in the shortest time possible, then create a Time Detective puzzle of their own. Ask, "Can you do all this in less than an hour?"
- ❑ Discuss different time measures invented by people before clocks. (e.g., sundials, water clocks, sand timers, candle clocks) Ask students, "How could you make one of these?"

Time Detective

O	J	A	N	U	A	R	Y	M	P
Y	R	U	O	H	M	A	Y	I	S
A	F	T	E	R	N	O	O	N	A
D	D	U	N	I	G	H	T	U	T
S	A	M	O	N	T	H	A	T	U
E	Y	N	L	S	U	M	M	E	R
N	O	V	E	M	B	E	R	J	D
D	B	A	M	T	S	U	G	U	A
E	G	N	I	R	P	S	O	L	Y
W	I	N	T	E	R	A	S	Y	B

1. JANUARY
2. TIME
3. AFTERNOON
4. SPRING
5. WEDNESDAY
6. SUMMER
7. HOUR
8. NIGHT
9. WINTER
10. SATURDAY
11. JULY
12. MONTH
13. DAY
14. AUTUMN
15. AUGUST
16. MINUTE
17. NOVEMBER
18. MAY

Time Detective (Answer Key)

O	J	A	N	U	A	R	Y	M	P
Y	R	U	O	H	M	A	Y	I	S
A	F	T	E	R	N	O	O	N	A
D	D	U	N	I	G	H	T	U	T
S	A	M	O	N	T	H	A	T	U
E	Y	N	L	S	U	M	M	E	R
N	O	V	E	M	B	E	R	J	D
D	B	A	M	T	S	U	G	U	A
E	G	N	I	R	P	S	O	L	Y
W	I	N	T	E	R	A	S	Y	B

1. JANUARY
2. TIME
3. AFTERNOON
4. SPRING
5. WEDNESDAY
6. SUMMER
7. HOUR
8. NIGHT
9. WINTER
10. SATURDAY
11. JULY
12. MONTH
13. DAY
14. AUTUMN
15. AUGUST
16. MINUTE
17. NOVEMBER
18. MAY

Things That Take an Hour	How I Timed It

Make an Analog Clock

Skills

- ❑ Tell time on the hour using an analog clock

Grouping

- ❑ whole class
- ❑ individuals

Materials

- ❑ examples of geared analog clocks
- ❑ paper, scissors, pencils, brads
- ❑ wristwatches (page 76), clear tape
- ❑ clocks (pages 77 and 78)
- ❑ matching game cards (pages 79 and 80)

Directions

- ❑ Ask students, "What sort of clocks do you use in your home?" (e.g., a wall clock, wristwatch, oven clock, computer) "In what ways are these clocks the same or different?" (e.g., Some have round faces. Some have just numbers.) Show examples of analog clocks. Ask, "What do you notice about analog clocks?" (e.g., They show time as groups of twelve hours. Time passing is shown as a circle with numbers in counting order from 1 to 12. The two hands show the hours and minutes ticking by.)
- ❑ Tell students that the direction in which the clock hands move is called *clockwise*. The hour hand is smaller. When it points to the hour, the longer minute hand is always pointing to the twelve. To tell the time in hours, you need to know whether it is before noon or after noon. Noon is when it is twelve o'clock in the middle of the day. (e.g., Later in the day when the hour hand is pointing exactly to three, you say, "It is three o'clock in the afternoon.") Discuss relevant o'clock events. (e.g., At eight o'clock, I get ready for bed.) Ask, "What time would it be if . . .? (e.g., What time would it be if I were eating my lunch?)
- ❑ Ask students, "How can you make a model clock?" (e.g., Trace around a plate on cardboard.) Have students write the numbers from 1–12 around the edge and cut it out. Then have them draw, cut out, and attach two hands with a brad in the center. Have students practice making specific o'clock times with a partner. You may also use the clock on page 92 for students to practice.

Variations

- ❑ Have students make a model wristwatch. Have them cut out and color a watch, then draw a secret o'clock time. Wrap the watch around the wrist and tape the ends together. Form a group of five students. Have them race to sort themselves in order from the earliest to the latest watch times.
- ❑ Have students practice identifying clock times using pages 77 and 78 or play the matching game on pages 79 and 80.

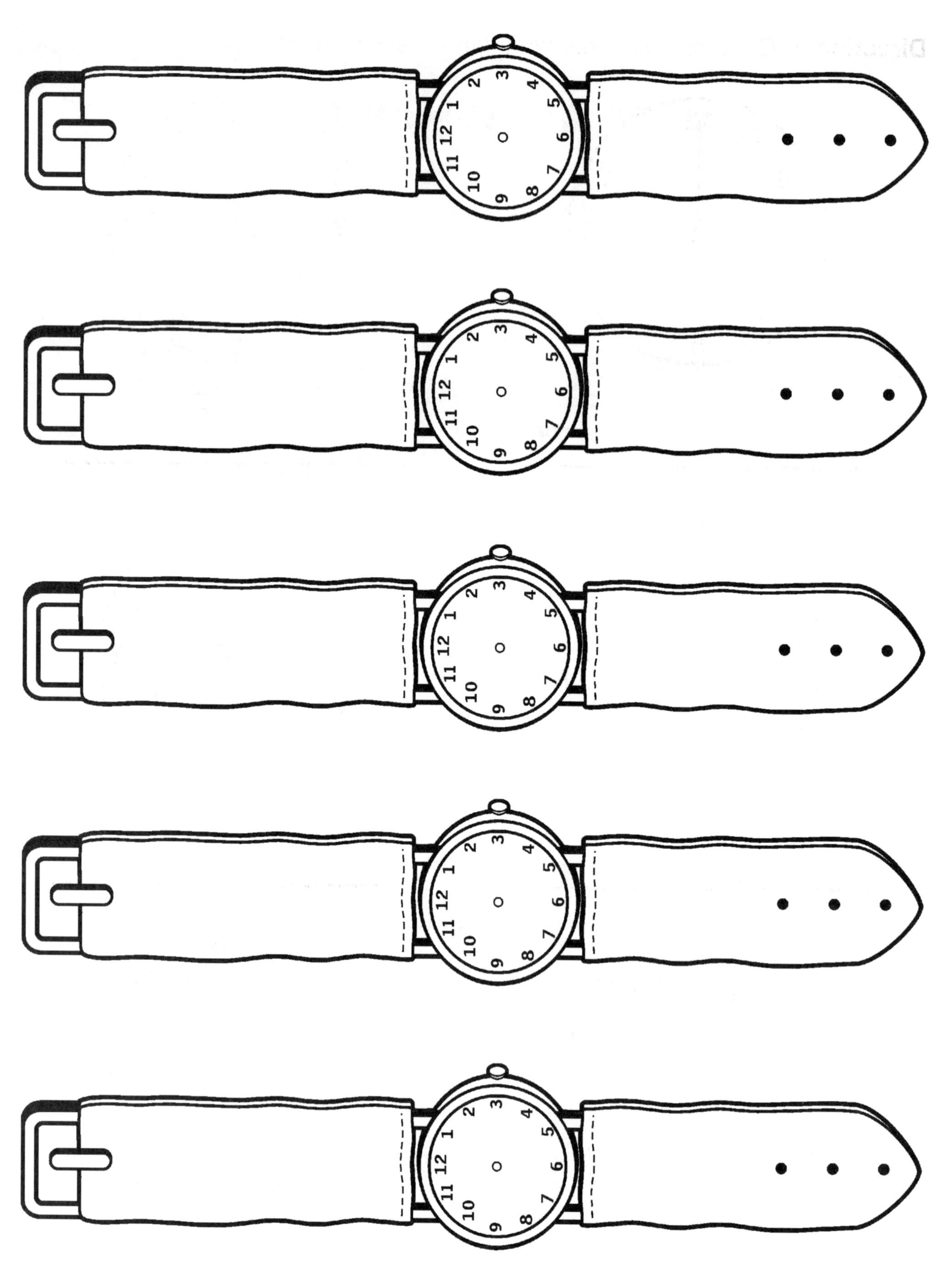

Directions: On each line, write what time is shown on the clock.

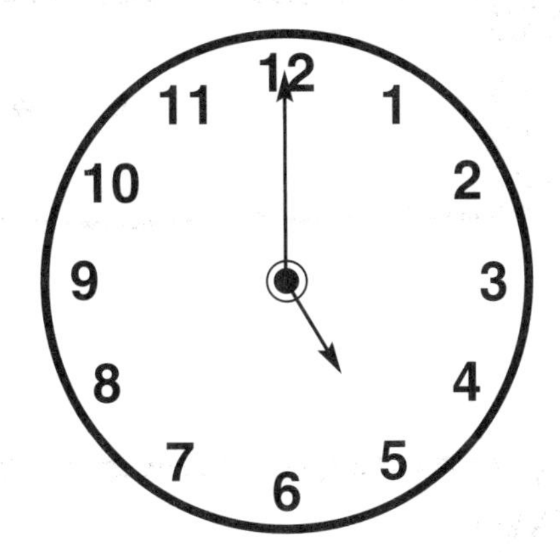

____

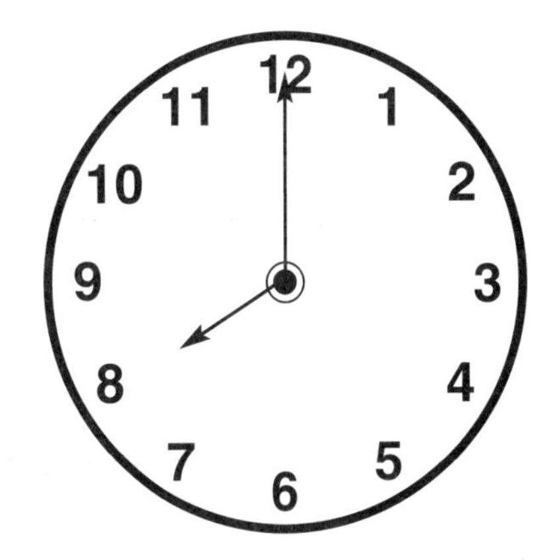

____

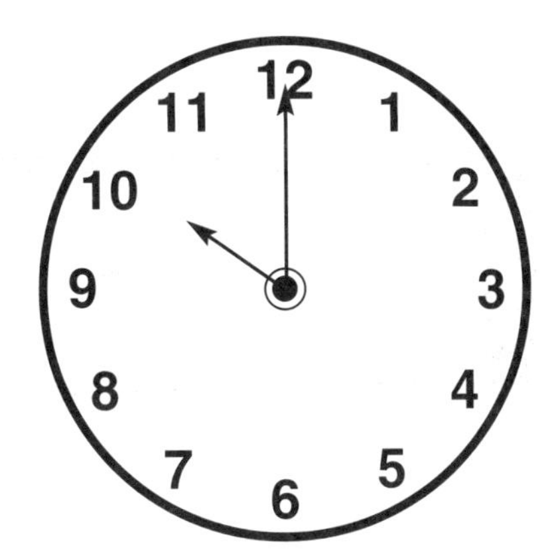

____

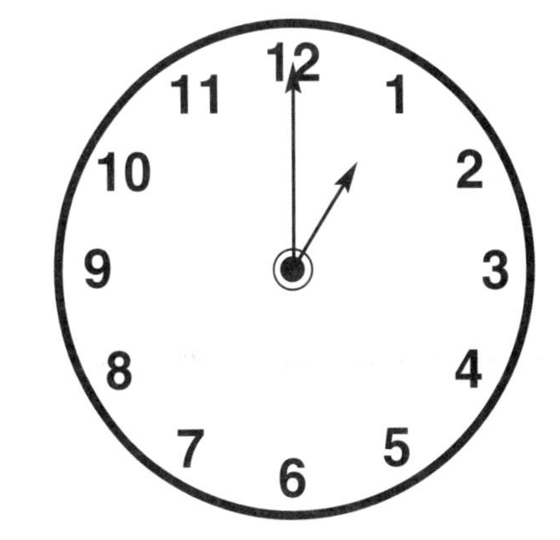

____

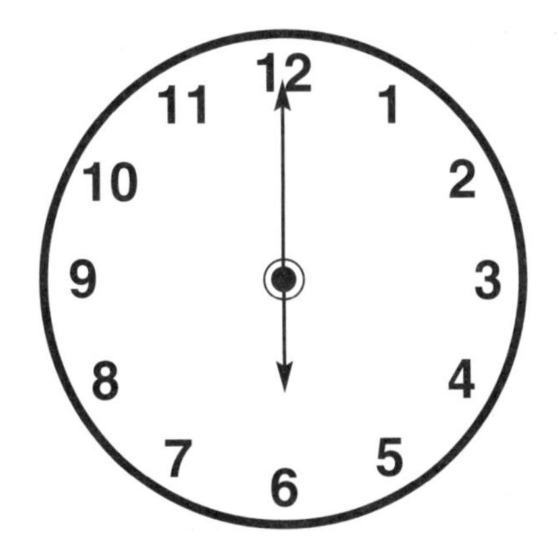

____

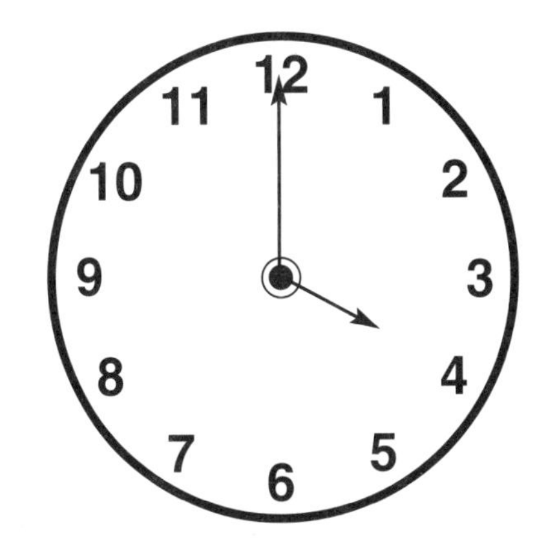

____

Directions: Draw the hands on the clock to match each time.

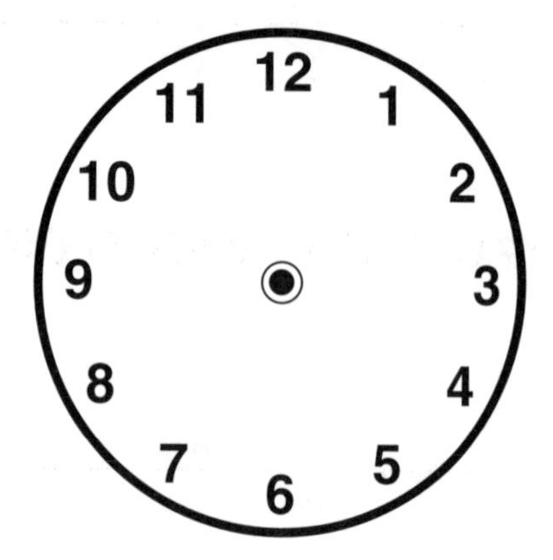

three o'clock

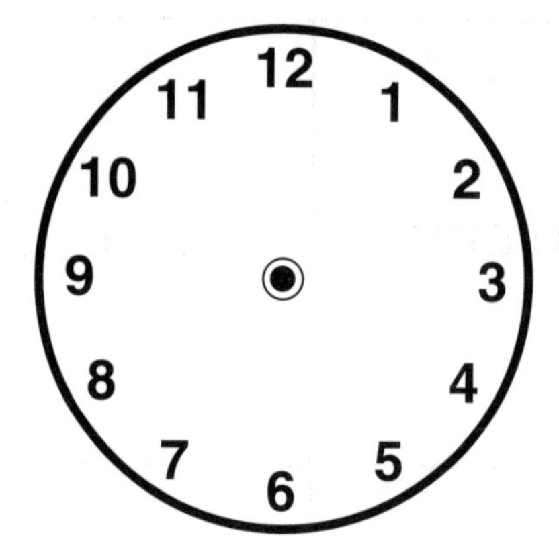

eleven o'clock

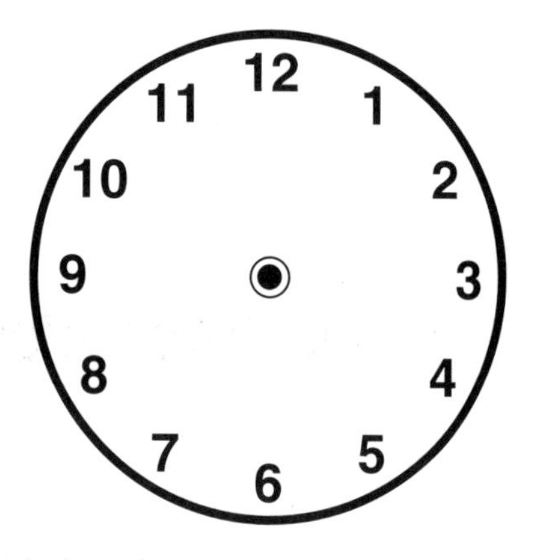

eight o'clock

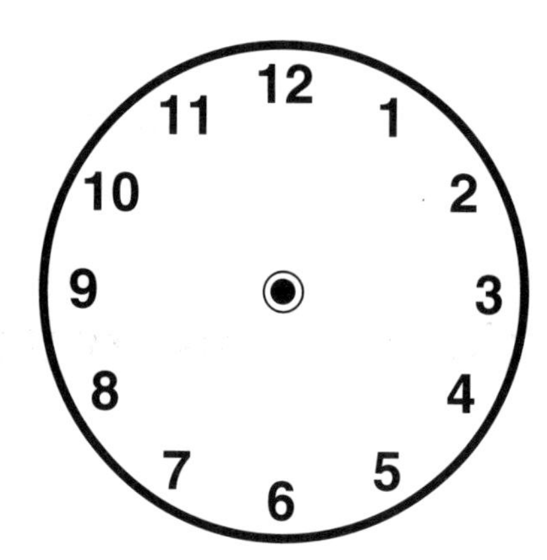

four o'clock

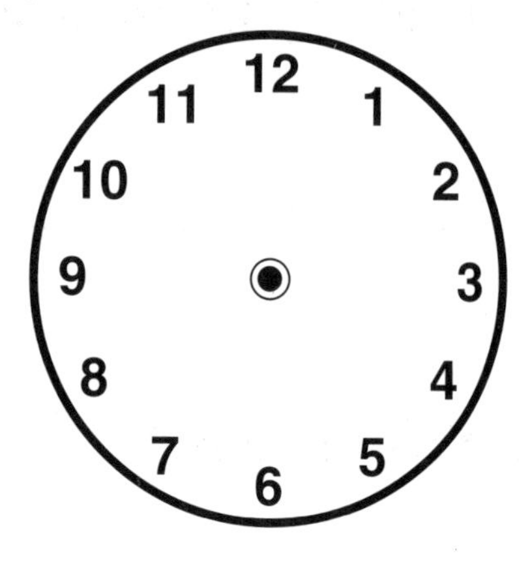

nine o'clock

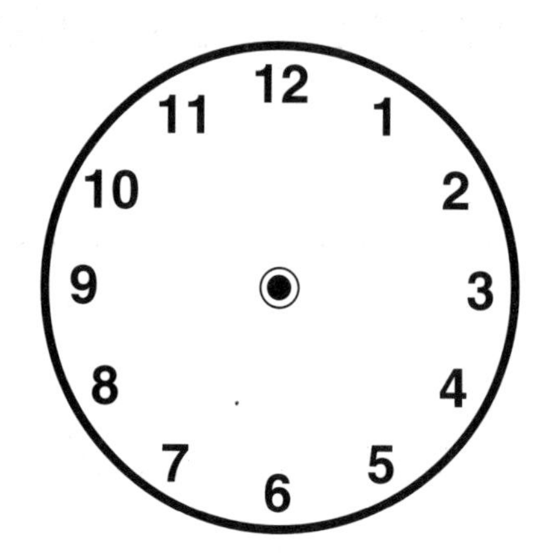

twelve o'clock

Directions: Cut out the cards below and match them to the clocks on page 80. Play a matching game.

one o'clock	two o'clock	three o'clock	four o'clock
five o'clock	six o'clock	seven o'clock	eight o'clock
nine o'clock	ten o'clock	eleven o'clock	three o'clock
five o'clock	six o'clock	seven o'clock	eight o'clock
nine o'clock	ten o'clock	eleven o'clock	twelve o'clock

Directions: Cut out the cards below and match them to the events on page 79. Play a matching game.

Make a Digital Clock

Skills

- ❑ Tell time on the hour using a digital clock

Grouping

- ❑ whole class
- ❑ individuals

Materials

- ❑ digital clocks
- ❑ digital time number strips (page 82)
- ❑ Time Challenge worksheet (page 83)
- ❑ digital time practice worksheets (pages 84 and 85)

Directions

- ❑ Ask students, "How do you tell the time using a digital clock?" (e.g., Time is divided into twelve hours like the analog clock. The hour is divided into sixty minutes. The clock face shows the hour numbers from 1–12 on the left and the minute numbers from 00 to 59 on the right.) Some digital clocks show all 24 hours on the left. The hour numbers go from 0 to 23. Tell students that when time reaches the next hour, the number on the left tells you the new hour. The number on the right is always 00. That tells them it is exactly on the hour and not a minute earlier or later.
- ❑ Discuss relevant o'clock events. (e.g., Sally was born at 4 o'clock in the morning.) Have students write the digital times for each event.
- ❑ Ask students, "How can you make your own digital clock?" (e.g., Decorate a small shoe box to look like a clock. Place matching digit cards to model a given time.) Or have them use the digital time number strips. Have students glue the tabs on the 5–0 strip to the 12–6 strip to make one long 12–0 hour strip on the left, and a 00–45 minute strip on the right. Have them thread these through cut-out holes in a small box or a piece of heavy paper and pull the strips around to match a given time. (See illustration at top.)

Variations

- ❑ Have students investigate the exact time they were born. Have them keep a record of everyone who was born at an o'clock time.
- ❑ Have students make a class picture book with magazine cut-outs and drawings showing different types of clocks.
- ❑ Have students complete the Time Challenge worksheet. Ask them to choose a time for each row and write it on all three clocks.
- ❑ Have students practice writing out digital times using the worksheets on pages 84 and 85.

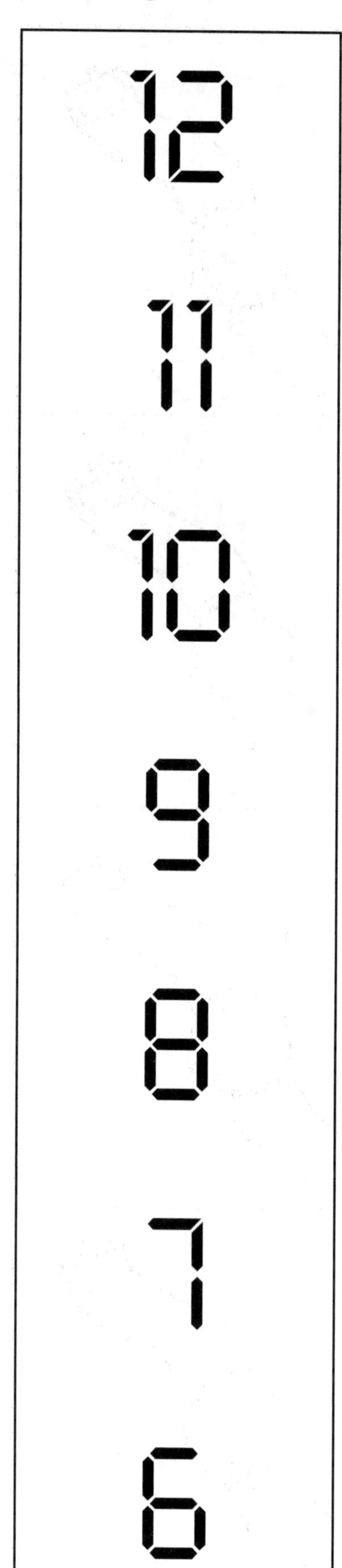

Tab

5

4

3

2

1

0

Tab

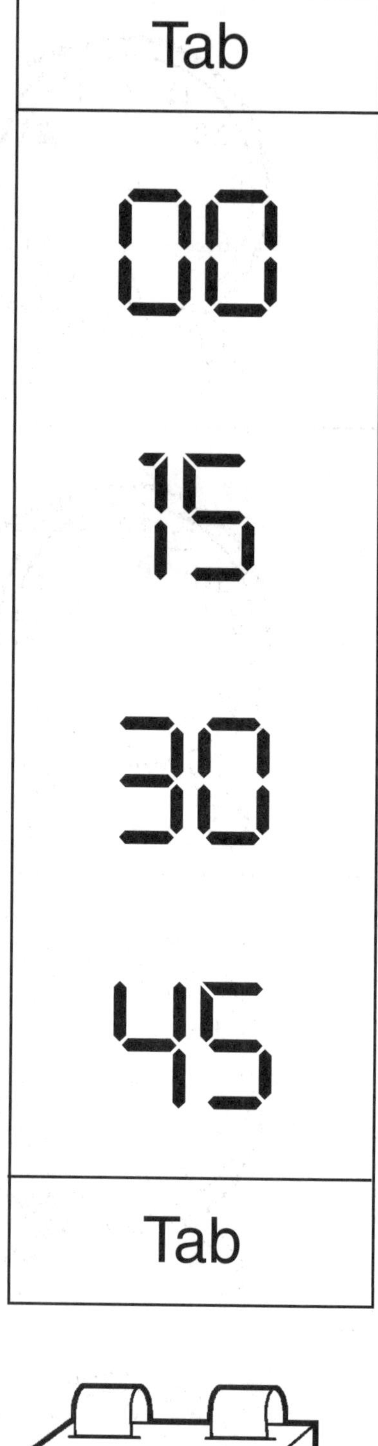

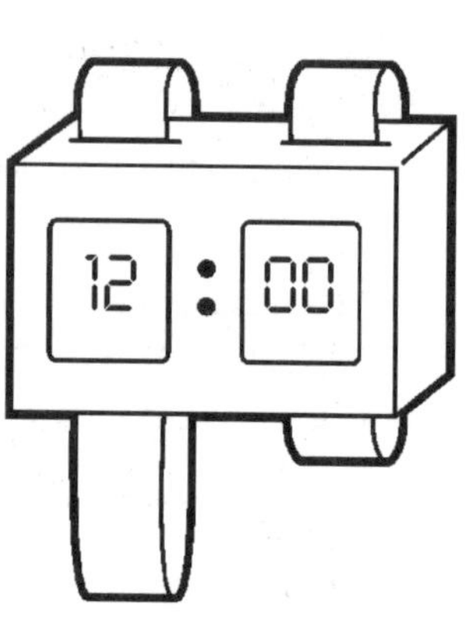

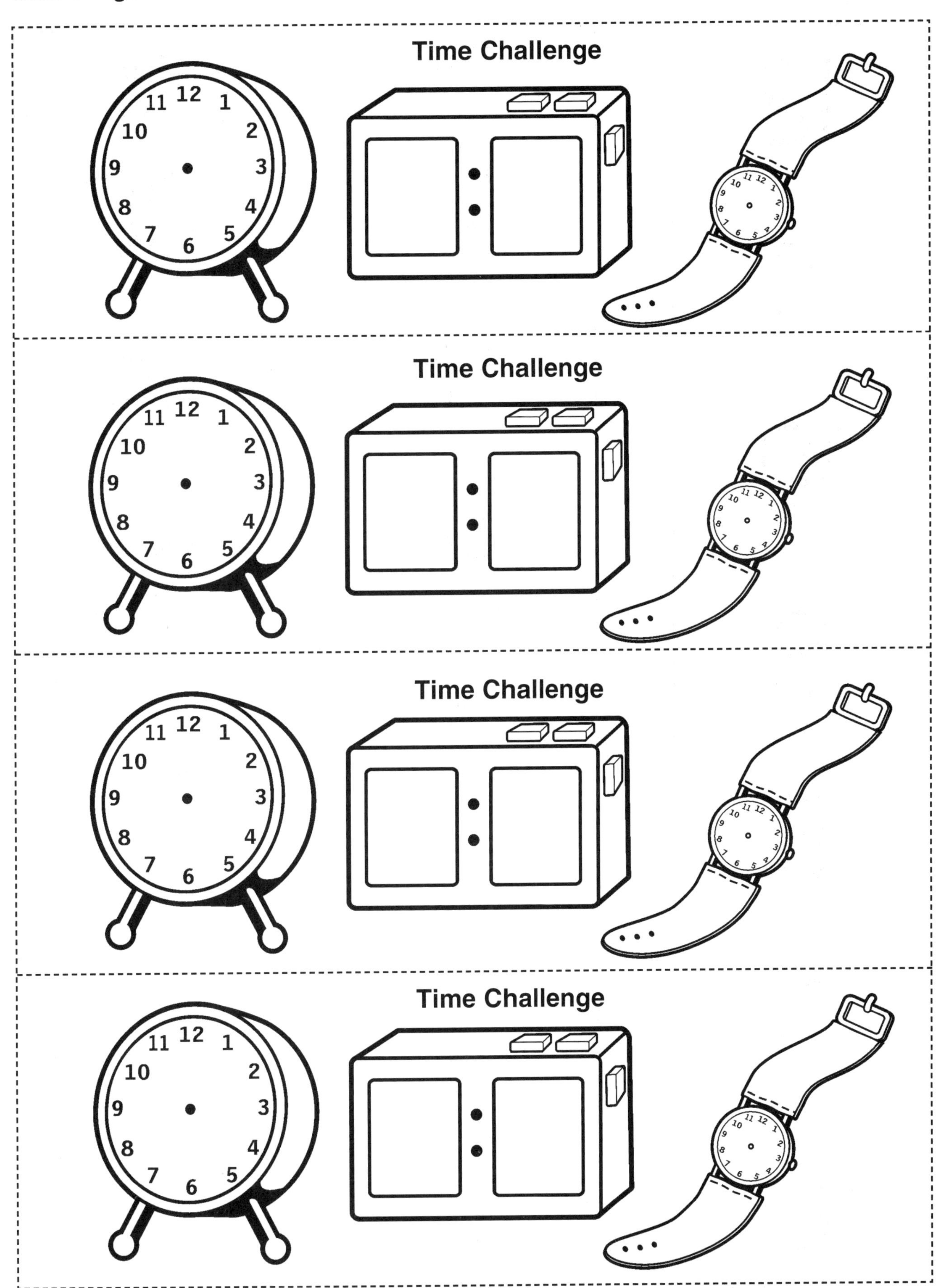
Time Challenge
12 1 2 3 4 5 6 7 8 9 10 11
Time Challenge
12 1 2 3 4 5 6 7 8 9 10 11
Time Challenge
12 1 2 3 4 5 6 7 8 9 10 11
Time Challenge
12 1 2 3 4 5 6 7 8 9 10 11

Directions: Write the time shown on each digital clock.

Directions: Write the time shown on each watch.

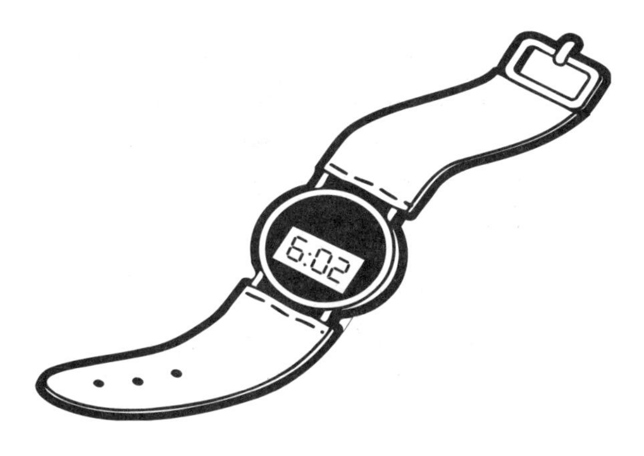

What Time Is It?

Skills

- ❑ Tell time on the hour using an analog clock
- ❑ Tell time on the hour using a digital clock

Grouping

- ❑ small groups

Materials

- ❑ a geared analog clock face (or a model with movable hands)
- ❑ a digital clock
- ❑ What Time Is It? cards (page 87)
- ❑ Special Events Log (page 88)

Directions

- ❑ Review with students how to tell the time on the hour using both an analog and a digital clock. (e.g., Notice the 12 is always at the top and the 6 is always at the bottom.)
- ❑ Ask students, "Could you still tell the time if there were no numbers on the clock face? Try it."
- ❑ Draw a large analog clock on the chalkboard. Draw 12 lines where the numbers would normally be. Have students practice drawing the clock hands to match a given time.

7 o'clock

- ❑ Have students practice telling the time on both analog and digital clocks.

- ❑ Have students lay the analog clock cards out on the table face up. Have students shuffle the digital clock cards and place them face down in a separate pile. When it is each student's turn, have him or her turn over the top digital card and say the time, then find the matching analog clock card.

Variations

- ❑ Students turn over an analog clock card and find the matching digital clock card.
- ❑ Students make a matching set of cards showing half-hour times.
- ❑ Students make a matching set of cards showing quarter to and quarter past the hour.
- ❑ Students make time diaries and record all the special events that occur in their days on each of the hours (and the half-hours, if applicable). Use the special events log on page 88.

1:00
7:00
2:00
8:00
3:00
9:00
4:00
10:00
5:00
11:00
6:00
12:00

Special Events Log

6 AM		**3 PM**	
7 AM		**4 PM**	
8 AM		**5 PM**	
9 AM		**6 PM**	
10 AM		**7 PM**	
11 AM		**8 PM**	
12 PM		**9 PM**	
1 PM		**10 PM**	
2 PM		**11 PM**	

Match My Time

Skills

- ❑ Tell time on the half-hour using an analog clock
- ❑ Tell time on the half-hour using a digital clock

Grouping

❑ pairs ❑ individuals

Materials

- ❑ analog and digital clocks
- ❑ Match My Time owl boards (page 90)
- ❑ Match My Time fact strips (page 91)
- ❑ scissors, pencils, glue

Directions

- ❑ Ask students, "How do you tell the time in hours and half-hours?" Explain how an analog clock shows half-hours. (e.g., The minute hand goes around a full circle in one hour, so after half an hour, the minute hand has only gone around half the circle. That means that it is pointing to the 6 at the bottom of the clock face. The hour hand is half way between the last hour and the next hour.)
- ❑ Ask students, "How do you read half-hours on a digital clock?" (e.g., An hour is represented by 60 minutes. So half an hour is shown by the number 30. The hour number is the same hour that has just passed.)
- ❑ Have students practice making and telling the time on the half-hour using analog and digital clocks.
- ❑ Play Match My Time. Have students color an owl card. Laminate *(optional)*, then have them carefully cut along the dashed lines at the top and bottom of the middle sections on each owl.
- ❑ Have students cut out two matching Digital A and Analog A fact strips. Have them thread the digital time strip from behind through the two slits on the left owl so that a fact appears on the owl's front. Then have them thread the matching analog strip in the same way through the two slits on the right owl. Have students glue the ends of each strip to create a continuous circular band. Have them find the match for each fact by pulling the strips through until a match is found. (See illustration at the top of this page.)

Variations

- ❑ Have students use the mixed Digital B/Analog B fact strips to practice mixed times.
- ❑ Have each students challenge a partner. (e.g., Ask him or her to show you amounts that are "one hour later than 2 o'clock" or "half an hour earlier than 7 o'clock")
- ❑ Have students use the blank Digital C/Analog C fact strips to write in their own time examples. Have them exchange strips with a friend and race to match all the facts.

Digital A	Analog A	Digital B	Analog B	Digital C	Analog C
6:30		4:00			
8:30		5:30			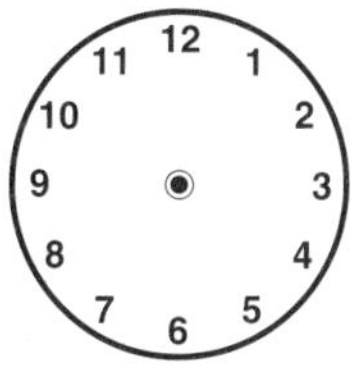
12:30		11:00			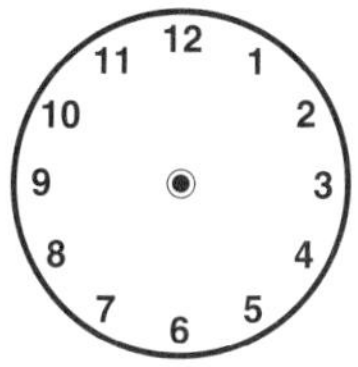
1:30		9:30			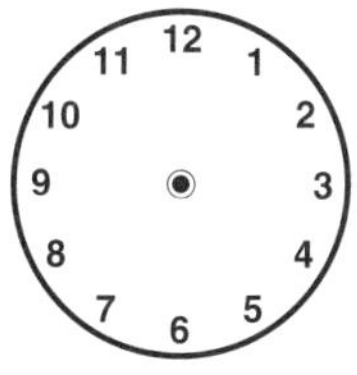
7:30		8:00			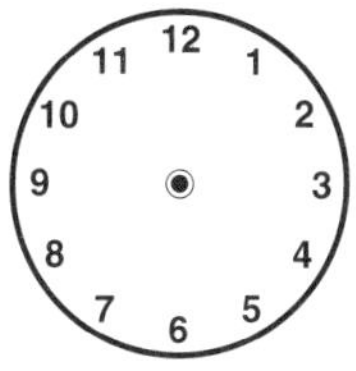
3:30		2:30	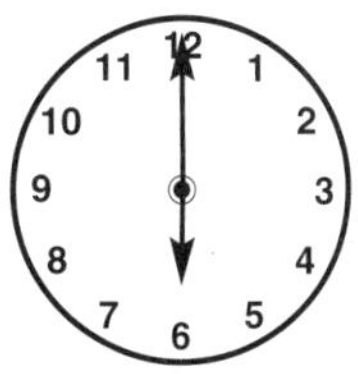		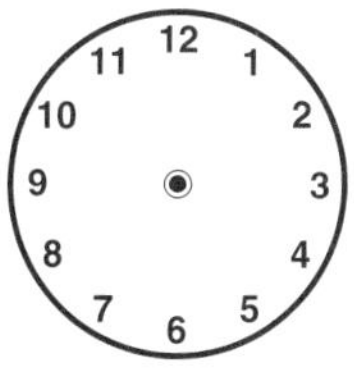
5:30		6:00			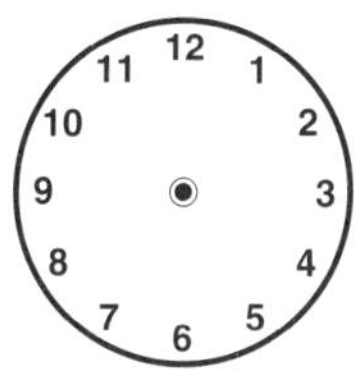
11:30		1:30			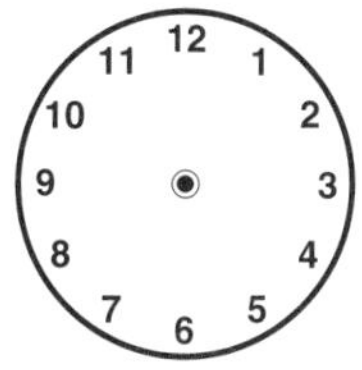

Clock

Directions: Copy or glue the clock onto cardstock. Color, cut out, and laminate for durability. Attach the hands of the clock with a brad.

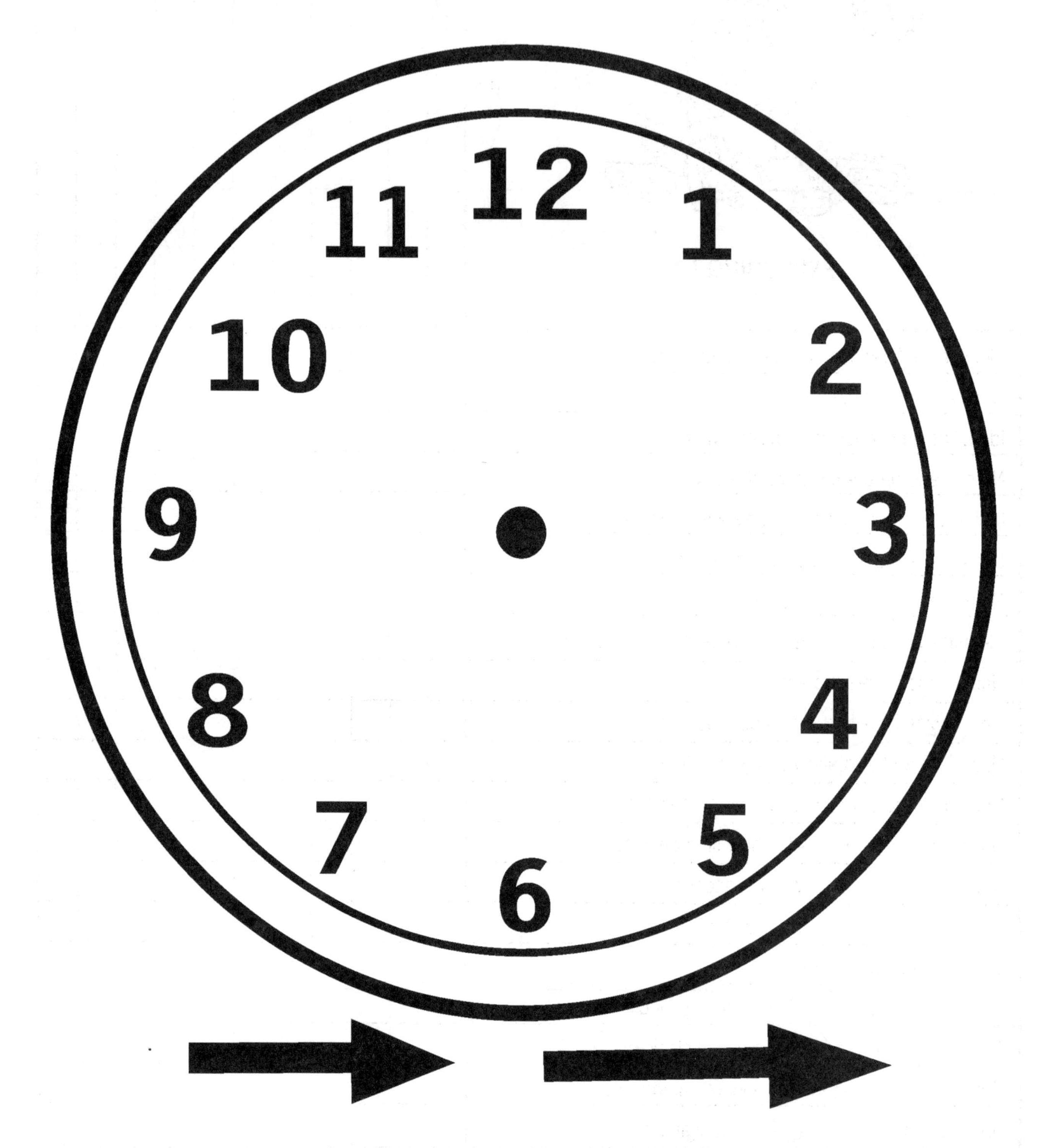

Skills Record Sheet

EXPLORING TIME

NAME											
Identifies differences between daytime and nighttime											
Uses everyday language related to day and night											
Identifies and sequences events within one day											
Names and orders the days of the week											
Identifies and sequences events within a week											
Uses the terms *yesterday, today,* and *tomorrow*											
Names and orders the months of the year											
Identifies and sequences special events within a year											
Names and orders the seasons											
Recognizes differences between months											
Uses a calendar to show a day and date											
Compares the duration of two or more events											
Uses informal time units to show time passing											
Demonstrates an awareness of seconds											
Demonstrates an awareness of minutes											
Demonstrates an awareness of an hour											
Tells time on the hour using an analog clock											
Tells time on the hour using a digital clock											
Tells time on the half-hour using an analog clock											
Tells time on the half-hour using a digital clock											

Sample Weekly Lesson Plan

STRAND Measurement

GRADE 1

SUBSTRAND Exploring Days of the Week

TERM 2 WEEK 7

OUTCOMES

- Name and order the days of the week
- Identify and sequence events within a week
- Use the terms *yesterday, today,* and *tomorrow*

LANGUAGE

- *weekend*
- *yesterday, today,* and *tomorrow*
- *sun* and *moon*, *day* and *night*
- *order, events*

RESOURCES

Turn-a-Week wheels (page 26)

Days-of-the-Week cards (page 29)

weather posters
large collection of magazines
Days of the Week (page 30)

Days of the Week (page 31)

Ways to Work with Weeks (pages 32–38)

MONDAY	TUESDAY	WEDNESDAY	THURSDAY	FRIDAY
What day of the week is it? • Whole class: - clues - names of the days - word list • Outdoor walk: Observation of sun and shadow movement • Recording: Chart special events	What day was yesterday? • Whole class: Discuss concepts of *yesterday* and *tomorrow* • Partner discussion: Cut and paste Days-of-the-Week cards in order	Weather • Whole class: - identify and list features of a hot day in summer - discuss seasonal variations - investigate posters/pic-tures - weather • As a group or individually, fill in Days of the Week (page 30)	Play Turn-a-Week (page 26) • Discuss special events (large and small) and work on page 31	• Distribute Days-of-the-Week cards. Have students find and line up in groups of seven—one for each day of the week • spinner matching game (pages 33 and 38) Note: At end of day, check weather predictions

Sample Weekly Lesson Plan

STRAND Measurement
GRADE 2

SUBSTRAND Time: Telling the time in hours
TERM 1 WEEK 8

OUTCOMES
- Demonstrate an awareness of hours
- Tell time on the hour using an analog clock
- Tell time on the hour using a digital clock
- Tell time on the half-hour using an analog clock
- Tell time on the half-hour using a digital clock

LANGUAGE
- "There are 60 minutes in an hour."
- "The hands go around in a clockwise direction."
- "It is now 10 o'clock . . . ."
- "That is an analog/digital clock."
- "Two hours earlier/later would be . . . half past three . . . , three thirty . . . ."

RESOURCES

MONDAY	TUESDAY	WEDNESDAY	THURSDAY	FRIDAY
candles, matches clock, stopwatch, alarm clock	geared analog clocks cardboard, brads, scissors	digital clocks small boxes digital strips (page 82) magazines, scissors worksheet (page 83)	geared analog, digital clocks paper/workbooks for time diaries wristwatches (page 76)	worksheets (pages 77 and 78)
How long is an hour? • Whole class: - What do you know about hours? - How can you measure an hour? - What lasts an hour? • Group activity: - List/sort events longer than, shorter than, same as one hour • Worksheet	Analog clocks • Whole class: - identify types of clocks - clockwise - how to tell the time in hours (Group A), half-hours (Group B) - observe position of hands • Make a human clock Group A - show *o'clock* Group B - show *half past* • List *o'clock, half past* events and match with clocks • Make model clocks - practice *o'clock, half past* times with a partner	Digital clocks • Whole class: - identify differences with digital clocks - how to tell digital time in hours, half-hours • List *o'clock, half past* events and match with clocks • Group activities: Group A - make model digital clock, record matching times on worksheet Group B - cut out, sort clock pictures, record 1, 2 hours later/earlier on worksheet	What time is it? • Whole class: - practice telling *o'clock, half past* times, position of hands - discuss analog clock face with no numbers, position of hands • Group activities: Group A - make matching *o'clock* time cards, time diaries Group B - make matching *half past* time cards, time diaries	Time revision • Challenge individual students to match given times on analog, digital clocks • Group A - show *o'clock* Group B - show h*alf past* • Make wristwatches - draw a secret time, form groups, and order times • Whole class challenge: What time would it be if . . .?

Weekly Lesson Plan

STRAND ____________________ SUBSTRAND ____________________

GRADE ____________________ TERM __________ WEEK __________

OUTCOMES

LANGUAGE

RESOURCES

MONDAY	TUESDAY	WEDNESDAY	THURSDAY	FRIDAY